D1095619

GPRS for Mobile Internet

For a listing of recent titles in the *Artech House Mobile Communications Series*, turn to the back of this book.

GPRS for Mobile Internet

Emmanuel Seurre
Patrick Savelli
Pierre-Jean Pietri

Artech House
Boston • London
www.artechhouse.com

Library of Congress Cataloging-in-Publication Data
Seurre, Emmanuel.
    GPRS for mobile Internet / Emmanuel Seurre, Patrick Savelli, Pierre-Jean Pietri.
      p. cm. (Artech House mobile communications series)
    Includes bibliographical references and index.
    ISBN 1-58053-600-X (alk. paper)
      1. Global system for mobile communications. 2. Wireless Internet—Standards.
    3. Mobile communication systems—Standards. I. Savelli, Patrick. II. Pietri, Pierre-Jean.
    III. Title.

    TK5103.483 .S48 2003
    621.39'81—dc21   2002038272

British Library Cataloguing in Publication Data
Seurre, Emmanuel.
    GPRS for mobile Internet. — (Artech House mobile communications series)
    1. Global system for mobile communications 2. General Packet Radio Service
    3. Wireless Internet
    I. Title II. Savelli, Patrick III. Pietri, Pierre-Jean
    621.3'845

    ISBN 1-58053-600-X

Cover design by Igor Valdman

© 2003 ARTECH HOUSE
685 Canton Street
Norwood, MA 02062

International Standard Book Number: 1-58053-600-X
Library of Congress Catalog Card Number: 2002038272

10 9 8 7 6 5 4 3 2 1

Contents

5   Radio Interface: RLC/MAC Layer   175

Acknowledgments

We would like to thank the following persons for their support during the redaction phase, and for their relevant comments on the manuscript: Jacques Achard, Solofoniaina Razafindrahaba, Samuel Rousselin, Jean-Louis Guillet, and Dominique Cyne.

Emmanuel Seurre
Patrick Savelli
Pierre-Jean Pietri
France
December 2002

1

Introduction to the GSM System

This chapter provides an overview of the GSM cellular system, with a focus on the radio interface. The purpose is not to give a detailed description of the many features supported by the system, but to summarize the elementary concepts of GSM, as an aid to reader comprehension of the subsequent chapters. In-depth presentation of the GSM system can be found in [1, 2].

1.1 Introduction

1.1.1 Birth of the GSM System

The first step in the history of GSM development was achieved back in 1979, at the World Administrative Radio Conference, with the reservation of the 900-MHz band. In 1982 at the *Conference of European Posts and Telegraphs* (CEPT) in Stockholm, the Groupe Spécial Mobile was created, to implement a common mobile phone service in Europe on this 900-MHz frequency band. Currently the acronym GSM stands for Global System for Mobile Communication; the term "global" was preferred due to the intended adoption of this standard in every continent of the world.

The proposed system had to meet certain criteria, such as:

- Good subjective speech quality (similar to the fixed network);
- Affordability of handheld terminals and service;
- Adaptability of handsets from country to country;

1

- Support for wide range of new services;
- Spectral efficiency improved with respect to the existing first-generation analog systems;
- Compatibility with the fixed voice network and the data networks such as ISDN;
- Security of transmissions.

Digital technology was chosen to ensure call quality.

The basic design of the system was set by 1987, after numerous discussions led to the choice of key elements such as the narrowband *time-division multiple access* (TDMA) scheme, or the modulation technique. In 1989 responsibility for the GSM was transferred to the *European Telecommunication Standards Institute* (ETSI). ETSI was asked by the EEC to unify European regulations in the telecommunications sector and in 1990 published phase I of the GSM system specifications (the phase 2 recommendations were published in 1995).

The first GSM handset prototypes were presented in Geneva for Telecom '91, where a GSM network was also set up. Commercial service had started by the end of 1991, and by 1993 there were 36 GSM networks in 22 countries. The system was standardized in Europe, but is now operational in more than 160 countries all over the world, and was adopted by 436 operators.

The growth of subscribers has been tremendous, reaching 500 million by May 2001.

1.1.2   The Standard Approach

The thousands of pages of GSM recommendations, designed by operators and infrastructure and mobile vendors, provide enough standardization to guarantee proper interworking between the components of the system. This is achieved by means of the functional and interface descriptions for each of the different entities.

The GSM today is still under improvement, with the definition of new features and evolution of existing features. This permanent evolution is reflected in the organization of the recommendations, first published as phase I, then phase II and phase II+, and now published with one release each year (releases 96, 97, 98, 99, and releases 4 and 5 in 2000 and 2001).

As stated, responsibility for the GSM specifications was carried by ETSI up to the end of 1999. During 2000, the responsibility of the GSM recommendations was transferred to the *Third Generation Partnership Project* (3GPP). This world organization was created to produce the third-generation mobile system specifications and technical reports. The partners have agreed to cooperate in the maintenance and development of GSM technical specifications and technical reports, including evolved radio access technologies [e.g., *General Packet Radio Service* (GPRS) and *Enhanced Data rates for Global Evolution* (EDGE)]. The structure and organization of the 3GPP is further described in Section 2.5.

1.2 General Concepts

1.2.1 Analog Versus Digital Telephony Systems

First-generation systems were analog. During the early 1980s these systems underwent rapid development in Europe. Although the NMT system was used by all the Nordic countries, and the TACS system in the United Kingdom and Italy, there was a variety of systems and no compatibility among them. Compared with these systems, the main advantages offered by GSM, which is the most important of the second-generation digital systems, are:

- Standardization;
- Capacity;
- Quality;
- Security.

Standardization guarantees compatibility among systems of different countries, allowing subscribers to use their own terminals in those countries that have adopted the digital standard. The lack of standardization in the first-generation system limited service to within the borders of a country. Mobility is improved, since roaming is no longer limited to areas covered by a certain system (see Section 1.2.6). Calls can be charged and handled using the same personal number even when the subscriber moves from one country to another.

Standardization also allows the operator to buy entities of the network from different vendors, since the functional elements of the network and the interfaces between these elements are standardized. This means that a mobile phone from any manufacturer is able to communicate with any network, even if this network is built with entities from different vendors. This leads

to a large economy of scale and results in cost reduction for both the operator and the subscriber. Furthermore, the phone cost is also reduced, because
as GSM is an international standard, produced quantities are greater and the
level of competition is high.

With respect to capacity, the use of the radio resource is much more
efficient in a digital system such as GSM than in an analog system. This
means that more users can be allocated in the same frequency bandwidth.
This is possible with the use of advanced digital techniques, such as voice
compression algorithms, channel coding, and multiple access techniques.
Note that capacity gains are also achieved with radio frequency reuse,
which had also been used in analog systems. Frequency reuse means that a
given carrier can be employed in different areas, as explained in Section
1.2.2.

The quality in digital transmission systems is better, thanks to the
channel coding schemes that increase the robustness in the face of noise and
disturbances such as interference caused by other users or other systems. The
quality improvement is also due to the improved control of the radio link,
and adaptations to propagation conditions, with advanced techniques such
as power control or frequency hopping. This will be explained in greater
detail in Section 1.5.6.3.

In terms of security, powerful authentication and encryption techniques for voice and data communications are enabled with GSM, which
guarantees protected access to the network, and confidentiality.

1.2.2 Cellular Telephony

In mobile radio systems, one of the most important factors is the frequency
spectrum. In order to make the best use of the bandwidth, the system is
designed by means of the division of the service area into neighboring zones,
or cells, which in theory have a hexagonal shape. Each cell has a *base transceiver station* (BTS), which to avoid interference operates on a set of radio
channels different from those of the adjacent cells. This division allows for
the use of the same frequencies in nonadjacent cells. A group of cells that as a
whole use the entire radio spectrum available to the operator is referred to as
a cluster. The shape of a cell is irregular, depending on the availability of a
spot for the BTS, the geography of the terrain, the propagation of the radio
signal in the presence of obstacles, and so on.

In dense urban areas, for instance, where the mobile telephony traffic is
important, the diameter of the cells is often reduced in order to increase

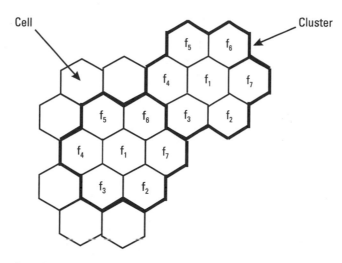

Figure 1.1 Example of a cell planning.

capacity. This is allowed since the same frequency channels are used in a smaller area. On the other hand, reducing the cell diameter leads to a decrease in the distance necessary to reuse the frequencies (that is, the distance between two cochannel cells), increasing cochannel interference. In order to minimize the level of interference, several techniques are used on the radio interface.

A basic example of cluster organization is shown in Figure 1.1. In this example, we see a reuse pattern for seven different frequencies, f_1 to f_7. These frequencies correspond to the beacon carrier of each cell, on which signaling information about the cell is broadcast (see Section 1.2.7). It can be seen from this figure that a given carrier can be reused in two separate geographical areas, as long as these areas are far enough from each other to reduce the effect of interference. With this technique of dividing the area in cells and clusters, the operator can increase the area it is able to cover with a limited frequency bandwidth.

1.2.3 Public Land Mobile Network

A *public land mobile network* (PLMN) is a network established for the purpose of providing land mobile telecommunications services to the public. It may be considered as an extension of a fixed network, such as the *Public Switched Telephone Network* (PSTN), or as an integral part of the PSTN.

1.2.4   Multiband Mobile Phones

Because of the increasing demand on the mobile networks, today the *mobile stations* (MSs) tend to be multiband. Indeed, to avoid network saturation in densely populated regions, mobile phones capable of supporting different frequency bands have been implemented, to allow for the user making communications in any area, at any time.

A dual-band phone can operate in two different frequency bands of the same technology, for instance in the 900-MHz and 1800-MHz frequency bands of the GSM system. Triple-band mobile phones have also come on the market, with the support of GSM-900 (900-MHz GSM band), DCS-1800 (1800-MHz GSM band), and PCS-1900 (1900-MHz GSM band), for example. Note that DCS-1800 and PCS-1900 are never deployed in the same country, and therefore this kind of phone can be used by travelers who want to have service coverage in a large number of countries.

1.2.5   SIM Card

One of the most interesting innovations of GSM is that the subscriber's data is not maintained in the mobile phone. Rather a "smart card," called a *subscriber identity module* (SIM) card, is used.

The SIM is inserted in the phone to allow the communications. A user may thus make telephone calls with a mobile phone that is not his own, or have several phones but only one contract. It is for example possible to use a SIM card in a different mobile when traveling to a country that has adopted the GSM on a different frequency band. A European can therefore rent a PCS1900 phone when traveling to the United States, while still using his own SIM card, and thus may receive or send calls. The SIM is used to keep names and phone numbers, in addition to those that are already kept in the phone's memory.

The card is also used for the protection of the subscriber, by means of a ciphering and authentication code.

1.2.6   Mobility

GSM is a cellular telephony system that supports mobility over a large area. Unlike cordless telephony systems, it provides location, roaming, and handover.

1.2.6.1 Location Area

The ability to locate a user is not supported in first-generation cellular systems. This means that when a mobile is called, the network has to broadcast the notification of this call in all the radio coverage. In GSM, however, *location areas* (LAs), which are groups of cells, are defined by the operator. The system is able to identify the LA in which the subscriber is located. This way, when a user receives a call, the notification (or paging) is only transmitted in this area. This is far more efficient, since the physical resource use is limited.

1.2.6.2 Roaming

In particular, the GSM system has the capability of international roaming, or the ability to make and receive phone calls to and from other nations as if one had never left home. This is possible because bilateral agreements have been signed between the different operators, to allow GSM mobile clients to take advantage of GSM services with the same subscription when traveling to different countries, as if they had a subscription to the local network. To allow this, the SIM card contains a list of the networks with which a roaming agreement exists.

When a user is "roaming" to a foreign country, the mobile phone automatically starts a search for a network stipulated on the SIM card list. The choice of a network is performed automatically, and if more than one network is given in the list, the choice is based on the order in which the operators appear. This order can be changed by the user. The home PLMN is the network in which the user has subscribed, while the visited PLMN often refers to the PLMN in which the user is roaming. When a user receives a call on a visited PLMN, the transfer of the call from the home PLMN to the visited PLMN is charged to the called user by his operator.

1.2.6.3 Handover

When the user is moving from one cell to the other during a call, the radio link between BTS 1 and the MS can be replaced by another link, between BTS 2 and the MS. The continuity of the call can be performed in a seamless way for the user. This is called handover. With respect to dual-band telephones, one interesting feature is called the dual-band handover. It allows the user in an area covered both by the GSM-900 and by the DCS-1800 frequency bands, for instance, to be able to transfer automatically from one system to the other in the middle of a call.

1.2.7 Beacon Channel

For each BTS of a GSM network, one frequency channel is used to broadcast general signaling information about this cell. This particular carrier frequency is called a beacon channel, and it is transmitted by the BTS with the maximum power used in the cell, so that every MS in the cell is able to receive it.

1.2.8 MS Idle Mode

When it is not in communication, but still powered on, the MS is said to be in idle mode. This means that it is in a low consumption mode, but synchronized to the network and able to receive or initiate calls.

1.3 GSM Services

In the specification of a telecommunication standard such as GSM, the first step is of course the definition of the services offered by the system. GSM is a digital cellular system designed to support a wide variety of services, depending on the user contract and the network and mobile equipment capabilities.

In GSM terminology, telecommunication services are divided into two broad categories:

- Bearer services are telecommunication services providing the capability of transmission of signals between access points [the *user-network interfaces* (UNIs) in ISDN]. For instance, synchronous dedicated packet data access is a bearer service.
- Teleservices are telecommunication services providing the complete capability, including terminal equipment functions, for communication between users according to protocols established by agreement between network operators.

In addition to these services, supplementary services are defined that modify or supplement a basic telecommunication service.

1.3.1 Bearer Services

There exist several categories of bearer services:

- *Unrestricted digital information* (UDI) is designed to offer a peer-to-peer digital link.

- The 3.1 kHz is external to the PLMN and provides a UDI service on the GSM network, interconnected with the ISDN or the PSTN by means of a modem.

- PAD allows an asynchronous connection to a *packet assembler/disassembler* (PAD). This enables the PLMN subscribers to access a *packet-switched public data network* (PSPDN).

- Packet enables a synchronous connection to access a PSPDN network and alternate speech and data, providing the capability to switch between voice and data during a call.

- Speech followed by data first provides a speech connection, and then allows to switch during the call for a data connection. The user cannot switch back to speech after the data portion.

1.3.2 Teleservices

In terms of application, teleservices correspond to the association of a particular terminal to one or several bearer services. They provide access to two kinds of applications:

- Between two compatible terminals;
- From an access point of the PLMN to a system including high-level functions, for example, a server.

Of course, the most basic teleservice supported by GSM is digital voice telephony, based on transmission of the digitally encoded voice over the radio. The voice service also includes emergency calls, for which the nearest emergency-service provider is notified by dialing three digits.

The other teleservices that are defined for a PLMN are:

- Data services, with data rates ranging from 2.4 Kbps to 14.4 Kbps. These services are based on circuit-switched technology. *Circuit switched* means that during the communication, a circuit is established between two entities for the transfer of data. The physical resource is used during the whole duration of the call.

- *Short message service* (SMS), which is a bidirectional service for short alphanumeric (up to 160 bytes) messages.

- Access to a voice message service.
- Fax transmission.

1.3.3 Supplementary Services

Supplementary services include several forms of call forward (such as call forwarding when the mobile subscriber is unreachable by the network), caller identification, call waiting, multiparty conversations, charging information, and call barring of outgoing or incoming calls. These call-barring features can be used for example when roaming in another country, if the user wants to limit the communication fees.

1.4   Network Architecture

The structure of a GSM network relies on several functional entities, which have been specified in terms of functions and interfaces. It involves three main subsystems, each containing functional units and interconnected with the others through a series of standard interfaces.

The main parts of a GSM network, as shown in Figure 1.2, are listed below. (In the figure, the lines between the entities represent the interfaces.)

- The MS, the handheld mobile terminal;
- The *base station subsystem* (BSS), which controls the radio link with the MS;
- The *network and switching subsystem* (NSS), which manages the function of connection switching to other fixed public network or mobile network subscribers, and handles the databases required for mobility management and for the subscriber data.

As can be seen from the figure, in a PLMN, the radio access network part (BSS) is logically separated from the core network (NSS), in order to ease the standardization of the different functions.

1.4.1   MS

The MS is made up of the *mobile equipment* (ME), and a SIM. It performs the following functions:

- Radio transmission and reception;
- Source and channel coding and decoding, modulation and demodulation;
- Audio functions (amplifiers, microphone, earphone);
- Protocols to handle radio functions: power control, frequency hopping, rules for access to the radio medium;
- Protocol to handle call control and mobility;
- Security algorithms (encryption techniques).

As mentioned, the SIM enables the user to have access to subscribed services irrespective of a specific terminal. The insertion of the SIM card into any GSM terminal allows the user to receive calls on that terminal, to make calls from that terminal, and to use the other subscribed services. The ME is identified with an *international mobile equipment identity* (IMEI).

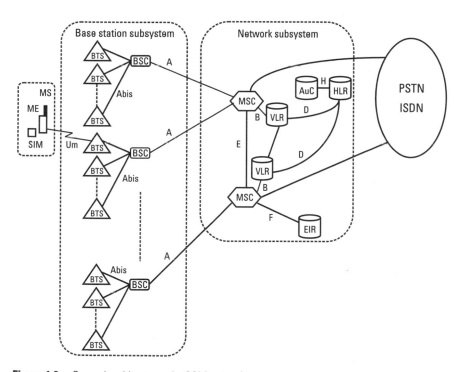

Figure 1.2 General architecture of a GSM network.

The SIM card contains, among other information, the *international mobile subscriber identity* (IMSI) used to identify the subscriber to the system, and a secret key for authentication. The IMEI and the IMSI are independent, thereby allowing personal mobility.

1.4.2   BSS

The BSS is composed of several *base station controllers* (BSCs) and BTS. These two elements communicate across the Abis interface.

The BTS contains the radio transceivers, responsible for the radio transmissions with the MS. This includes the following functions:

- Modulation and demodulation;
- Channel coding and decoding;
- Encryption process;
- RF transmit and receive circuits (power control, frequency hopping, management of antenna diversity, discontinuous transmission).

Several types of BTS exist: the normal BTS, the micro BTS, and the pico BTS. The micro BTS is different from a normal BTS in two ways. First, the range requirements are reduced, and the close proximity requirements are more stringent. Second, the micro BTS is required to be small and affordable in order to allow external street deployment in large numbers. The pico BTS is an extension of the micro BTS concept to the indoor environments. The RF performances of these different BTSs are slightly different.

The BSC manages the radio resources for one or more BTSs. It handles the management of the radio resource, and as such it controls the following functions: allocation and release of radio channels, frequency hopping, power control algorithms, handover management, choice of the encryption algorithm, and monitoring of the radio link.

1.4.3   Network Subsystem

The *mobile services switching center* (MSC) is the central part of the *network subsystem* (NSS). It is responsible for the switching of calls between the mobile users (between different BSCs or toward another MSC) and between mobile and fixed network users. It manages outgoing and incoming calls from various types of networks, such as PSTN, ISDN, and PDN. It also handles the functionality required for the registration and authentication of a user, and the mobility operations. This includes location updating, inter-MSC handovers, and call routing.

The BSS communicates with the MSC across the A interface.

Associated with the MSC, two databases, the *home location register* (HLR) and the *visitor location register* (VLR), provide the call-routing and roaming capabilities. The HLR contains all the administrative information related to the registered subscribers within the GSM network. This includes the IMSI, which unequivocally identifies the subscriber within any GSM network, the *MS ISDN number* (MSISDN), and the list of services subscribed by the user (such as voice, data service). The HLR also stores the current location of the MS, by means of the address of the VLR in which it is registered.

The VLR temporarily keeps the administrative data of the subscribers that are currently located in a given geographical area under its control.

Each functional entity may be implemented as an independent unit, but most of the time, the VLR is colocated with the MSC, so that the geographical area controlled by the MSC corresponds to that controlled by the VLR. The MSC contains no information about particular MSs, but rather, the information is stored in the location registers.

Two other registers are used for authentication and security purposes:

- The *equipment identity register* (EIR) is a database that contains a list of all valid ME on the network, where each MS is identified by its IMEI. An IMEI is marked as invalid if it has been reported stolen.

- The *authentication center* (AuC) is a protected database that contains a copy of the secret key stored in each subscriber's SIM card, for authentication and encryption over the radio channel. The AuC verifies if a legitimate subscriber has requested a service. It provides the codes for both authentication and encryption to avoid undesired violations of the system by third parties.

Two other important entities of the NSS are the *operations and maintenance center* (OMC) and the *network management center* (NMC). These entities perform the functions relative to the *network management* (NM), such as the configuration of the system (locally or remotely), maintenance and tests of the pieces of equipment, billing, statistics on the performance, and gathering of all information related to subscriber traffic necessary for invoicing and administration of subscribers.

1.5   Radio Interface

1.5.1   General Characteristics

Currently, there are several types of networks in the world using the GSM standard, but at different frequencies.

- The GSM-900 is the most common in Europe and the rest of the world. Its extension is E-GSM.
- The DCS-1800 operates in the 1,800-MHz band and is used mainly in Europe, usually to cover urban areas. It was also introduced to avoid saturation problems with the GSM-900.
- The PCS-1900 is used primarily in North America.
- The GSM-850 is under development in America.
- The GSM-400 is intended for deployment in Scandinavian countries in the band previously used for the analog *Nordic Mobile Telephony* (NMT) system.

The system is based on *frequency-division duplex* (FDD), which means that the uplink (radio link from the mobile to the network—that is, mobile transmit, base receive), and downlink (from the network to the mobile—that is, base transmit, mobile receive) are transmitted on different frequency bands. For instance, in the 900-MHz E-GSM band, the block 880–915 MHz is used for transmission from mobiles to network, and the

Table 1.1
GSM System Frequency Bands

		Uplink Band	Downlink Band
GSM-900		890–915 MHz	935–960 MHz
E-GSM-900		880–915 MHz	925–960 MHz
DCS-1800		1,710–1,785 MHz	1,805–1,880 MHz
PCS-1900		1,850–1,910 MHz	1,930–1,990 MHz
GSM-400	GSM-450	450,4–457,6 MHz	460.4–467.6 MHz
	GSM-480	478.8–486 MHz	488.8–496 MHz
GSM-850		824–849 MHz	869–894 MHz

block 925–960 MHz is used for the transmission from network to mobiles. Table 1.1 gives a summary of uplink and downlink frequency bands for the different GSM systems.

Operators may implement networks that operate on a combination of the frequency bands listed above to support multiband mobile terminals.

There are different ways of sharing the physical resource among all the users in a radio system, and this is called the multiple-access method. The multiple-access scheme defines how simultaneous communications share the GSM radio spectrum. The various multiple-access techniques in use in radio systems are *frequency-division multiple access* (FDMA), TDMA, and *code-division multiple access* (CDMA). GSM is based on both FDMA and TDMA techniques (see Figure 1.3).

FDMA consists in dividing the frequency band of the system into several channels. In GSM, each RF channel has a bandwidth of 200 kHz, which is used to convey radio modulated signals, or carriers. Each pair of uplink/downlink channels is called an *absolute radio frequency channel* (ARFC) and is assigned an *ARFC number* (ARFCN). The mapping of each ARFCN on the corresponding carrier frequency is given in [3].

TDMA is the division of the time into intervals: within a frequency channel, the time is divided into time slots. This division allows several users

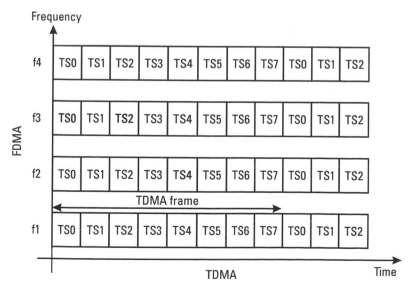

Figure 1.3   TDMA and FDMA.

(eight) to be multiplexed on the same carrier frequency, each user being assigned a single time slot. A packet of data information, called a burst, is transmitted during a time slot. The succession of eight time slots is called a TDMA frame, and each time slot belonging to a TDMA frame is identified by a *time slot number* (TN), from 0 to 7.

1.5.2 Logical Channels

The association of a radio frequency channel and a time slot—the pair ARFCN and TN—uniquely defines a physical channel on both the uplink and the downlink.

On top of the physical channels, logical channels ar mapped to convey the information of voice, data, and signaling. This signaling information is used for setting up a call, or to adapt the link to rapidly changing radio conditions, or to manage handovers, to give a few examples. Logical channels can be seen as pipes, each one used for a different purpose by the higher layers of the system.

Two types of logical channels exist, traffic channels and control channels. Among the control channels, according to their functions, four classes are defined: broadcast, dedicated, common, and associated. A broadcast channel is used by the network (in downlink only) to send general information to the MSs. A channel is said to be dedicated if only one MS can transmit or receive in the ARFCN-TN defining this channel, and common if it carries information for several mobiles. An associated control channel is allocated to one mobile, in addition to a dedicated channel, and carries signaling for the operation of this channel.

The broadcast channels are transmitted on the beacon carrier frequency presented in Section 1.2.7. The purposes of the beacon are:

- To allow a synchronization in time and frequency of the MSs to the BTS. This synchronization is needed by the MS to access the services of a cell. The frequency and time synchronization procedures that are performed by the mobile are explained in Section 1.5.7.

- To help the mobile in estimating the quality of the link during a communication, by measurements on the received signal from the BTS it is transmitting to, and from the other BTSs of the geographical area. These measurements are used by the network to determine when a handover is necessary, and to which BTS this handover should apply.

- To help the mobile in the selection of a cell when it is in idle mode (that is, not in communication, but still synchronized to the system and able to receive an incoming call or to initiate a call). This selection is performed on the basis of the received power measurements made on the adjacent cells' beacon channels.

- To access the general parameters of the cell needed for the procedures applied by the MS, or general information concerning the cell, such as its identification, the beacon frequencies of the surrounding cells, or the option supported by the cell (services).

To allow these various operations, the logical channels transmitted on the beacon are:

- The *broadcast control channel* (BCCH), which continually broadcasts, on the downlink, general information on the cell, including base station identity, frequency allocations, and frequency-hopping sequences. The information is transmitted within *system information* (SI) blocks, which can be of different types according to the information that is carried out. The frequency with which an SI is retransmitted on the BCCH varies with the type of information.

- The *frequency control channel* (FCCH), used by the MS to adjust its *local oscillator* (LO) to the BTS oscillator, in order to have a frequency synchronization between the MS and the BTS.

- The *synchronization channel* (SCH), used by the MS to synchronize in time with the BTS, and to identify the cell.

As listed below, four channels comprise the *common control channels* (CCCH). Among these, the first three are used for the MS-initiated call or for call paging (notification of an incoming call toward the MS):

- The *random access channel* (RACH) is used for the MS access requests to the network, for the establishment of a call, based on a slotted aloha method.

- The *paging channel* (PCH) is defined to inform the MS of an incoming call.

- The *access grant channel* (AGCH) is used to allocate some physical resource to a mobile for signaling, following a request on the RACH.

- The *cell broadcast channel* (CBCH) may be used to broadcast specific news to the mobiles of a cell.

The dedicated control channels are:

- The *stand-alone dedicated control channel* (SDCCH), utilized for registration, authentication, call setup, and location updating.
- The *slow associated control channel* (SACCH), which carries signaling for the TCH or SDCCH with which it corresponds. The information that is transmitted on this channel concerns the *radio link control* (RLC), such as the power control on the corresponding TCH or SDCCH, or the time synchronization between the MS and the BTS.
- The *fast associated control channel* (FACCH), carries the signaling that must be sent by the network to the MS to notify that a handover is occurring.

The TCHs can be of several types, according to the service that is accessed by the subscribers: voice or data, with various possible data rates.

Table 1.2 summarizes the purpose of the different logical channels. In this table, UL stands for uplink, and DL for downlink.

1.5.3 Mapping of Logical Channels onto Physical Channels

1.5.3.1 TDMA Time Structure

The basic time unit is the time slot. Its duration is 576.9 μs = 15/26 ms, or 156.25 symbol periods (a symbol period is 48/13 μs). The piece of information transmitted during a time slot is called a burst. As we saw in Section 1.5.1, the GSM multiple access scheme is TDMA, with eight time slots per carrier. A sequence of eight time slots is called a TDMA frame, and has a duration of 4.615 ms. The time slots of a TDMA frame are numbered from 0 to 7, as shown in Figure 1.4. Note that the beginning and end of TDMA frames in uplink and downlink are shifted in time: Time slot number 0 on the uplink corresponds to time slot 3 in the downlink. This allows some time for the mobile to switch from one frequency to the other.

As seen earlier, a physical channel is defined as a sequence of TDMA frames, a time slot number (from 0 to 7) and a frequency. It is bidirectional,

Table 1.2
The Logical Channels and Their Purpose

	Logical Channel	Abbreviation	Uplink/ Downlink	Task
Broadcast channel (BCH)	Broadcast control channel	BCCH	DL	System Information broadcast
	Frequency correction channel	FCCH	DL	Cell frequency synchronization
	Synchronization channel	SCH	DL	Cell time synchronization and identification
Common control channel (CCCH)	Paging channel	PCH	DL	MS paging
	Random access channel	RACH	UL	MS random access
	Access grant channel	AGCH	DL	Resource allocation
	Cell broadcast channel	CBCH	DL	Short messages broadcast
Dedicated control channel	Standalone dedicated control channel	SDCCH	UL/DL	General signaling
	Slow associated control channel	SACCH	UL/DL	Signaling associated with the TCH
	Fast associated control channel	FACCH	UL/DL	Handover signaling
Traffic channel (TCH)	Full speech	TCH/FS	UL/DL	Full-rate voice channel
	Half rate	TCH/HS	UL/DL	Half-rate voice channel
	2.4 Kbps, 4.8 Kbps, 9.6 Kbps, and 14.4 Kbps full-rate data channels	TCH/F2.4 TCH/F4.8 TCH/F9.6 TCH/F14.4	UL/DL	Full-rate data channels
	2.4-Kbps- and 4.8-Kbps-rate data channels	TCH/H2.4 TCH/H4.8	UL/DL	Half-rate data channels

with the same TN in uplink and in downlink. In order to support cryptographic mechanisms, a long time-structure has been defined. It is called a *hyperframe* and has a duration of 3 hours, 28 minutes, 53 seconds, and 760 ms (or 12,533.76 seconds). The TDMA frames are numbered within the hyperframe. The numbering is done with the TDMA *frame number* (FN) from 0 to 2,715,647.

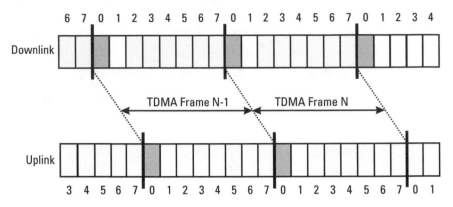

Figure 1.4   Slot numbering within the TDMA frame.

One hyperframe is subdivided into 2,048 superframes, which have a duration of 6.12 seconds. The superframe is itself subdivided into multiframes. In GSM, there are two types of multiframes defined, containing 26 or 51 TDMA frames.

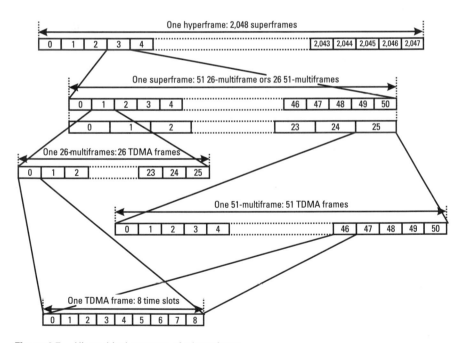

Figure 1.5   Hierarchical structure of a hyperframe.

The 26 multiframe has a duration of 120 ms, and comprises 26 TDMA frames This multiframe is used to carry TCH, SACCH, and FACCH. The 51 multiframe is made up of 51 TDMA frames. Its duration is 235.4 ms (3,060/13 ms). This multiframe is used to carry BCH, CCCH, and SDCCH (with its associated SACCH). Note that a superframe is composed of twenty-six 51-multiframes, or of fifty-one 26-multiframes. This hierarchical time structure is summarized in Figure 1.5.

1.5.3.2 Mapping of the TCH and SACCH on the 26-Multiframe

The TCHs are bidirectional channels mapped onto the 26-multiframe. Two types of channels must be distinguished: full-rate and half-rate channels, and therefore two different mappings of the TCH on the multiframe are possible:

- A full-rate traffic channel (TCH/FS, for full speech) uses one time slot per TDMA frame, for each frame of the multiframe, except the frames 12 and 25 (see Figure 1.6). The TDMA frame 12 is used to carry the SACCH/FS, and the TDMA frame 25 is an idle frame, which means that no channel is transmitted during this entire TDMA frame.

- A half-rate traffic channel (TCH/HS) uses one time slot every two TDMA frames, due to the fact that it carries data from a half-rate voice coder. As shown in Figure 1.7, two half-rate channels can be mapped on the same time slot, one using TDMA frames 0, 2, 4, 6, 8, 10, 13, 15, 17, 19, 21, and 23 and the other one using frames 1, 3, 5, 7, 9, 11, 14, 16, 18, 20, 22, and 24. The SACCH/HS channel

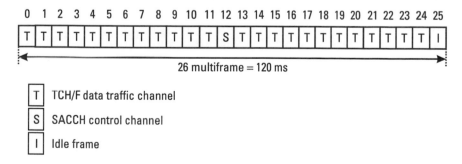

Figure 1.6   Mapping of a TCH/FS and SACCH/FS on the 26-multiframe.

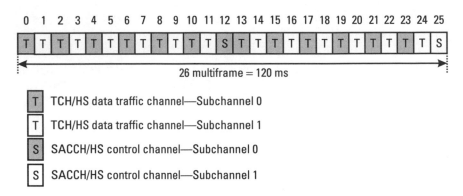

Figure 1.7　Mapping of a TCH/HS and SACCH/HS on the 26-multiframe.

associated with the first TCH subchannel is transported on TDMA frame 12, and the SACCH/HS associated with the second subchannel is on time slot 25.

1.5.3.3　Mapping of the FACCH on the 26-Multiframe

The FACCH is associated with a TCH, and is required to support the high-speed signaling needed during call establishment, subscriber authentication, and handover management. The occurrence of the FACCH is not fixed in the multiframe, as it is for the SACCH. Rather, the FACCH occurs on a TDMA frame that is reserved for a TCH. The multiplexing of TCH and FACCH is possible by means of the frame stealing. This means that a speech frame carried over a TCH can be replaced by a FACCH frame. This is signaled to the receiver by means of stealing flags, as described in Section 1.5.4. This principle allows for a fast signaling channel without significant loss on the quality of speech, if it is not performed too often.

1.5.3.4　Mapping of the SDCCH on the 51-Multiframe

The SDCCH is a signaling channel that carries the higher layers of control information. A SACCH is associated with a SDCCH.

Two different multiplexings on the 51 multiframe are possible:

- *SDCCH with SACCH alone.* An SDCCH channel is mapped on four TDMA frames of a 51 multiframe. As a result, eight SDCCH channels, dedicated to eight different MSs are mapped onto a 51 multiframe. The TDMA frames not occupied by an SDCCH are used by the eight associated SACCH channels. The mapping of these associated channels is performed on two consecutive 51 multiframes, as shown in Figure 1.8(b).

* *SDCCH with SACCH multiplexed together with CCCH, BCCH, SCH. and FCCH.* This case is described in the next subsection and in Figure 1.8(c).

Note that the frame-stealing concept, used on the TCH to allocate FACCH frames, is not in use in the case of an SDCCH. This is due to the fact that sufficiently fast signaling can be transmitted over an SDCCH to carry on a handover procedure.

1.5.3.5   Mapping of the Broadcast Channels and CCCH on the 51 Multiframe

The broadcast channels and the CCCH (i.e., PCH, RACH, AGCH) are all multiplexed on the 51 multiframe, on the beacon carrier frequency. The FCCH is sent on downlink time slot 0, on the TDMA frames 0, 10, 20, 30, and 40 of the 51 multiframe. The SCH is also mapped on slot 0, in the TDMA frames immediately following an FCCH frame (i.e., frames 1, 11, 21, 31, 41).

The BCCH is associated with time slot 0 as well, in the TDMA frames that are not occupied by the FCCH and the SCH, and can also be mapped on time slots 2, 4, and 6 in some configurations.

The AGCH and PCH are dynamically multiplexed on the multiframe according to the network load. They are mapped on time slot 0, together with the FCCH, SCH, and BCCH, and optionally on time slots 2, 4, and 6 (also possibly used for the BCCH). Note that the BCCH and CCCH are multiplexed dynamically. The exact mapping is conveyed to the MS by means of SI blocks, sent over the BCCH.

One of these broadcast parameters also indicates whether or not the CCCH are combined with SDCCH and SACCH onto the same basic physical channel (the second type of mapping of SDCCH/SACCH presented in Section 1.5.3.4).

The mapping of the RACH channel is simple: every uplink slot corresponding to a downlink FCCH, SCH, BCCH, PCH, and AGCH can be used for a RACH. If the SDCCH and SACCH are multiplexed with the CCCH, the number of available random-access channel blocks are reduced. Figure 1.8(a, c) shows possible configurations with the multiplexing BCCH/CCCH and BCCH/CCCH/SDCCH/SACCH.

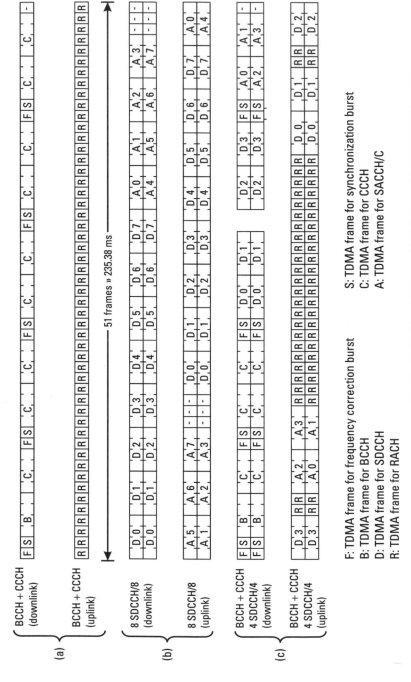

Figure 1.8 Channel associations on the 51-multiframe. (a) BCCH + CCCH, (b) 8 SDCCH/8, and (c) BCCH + CCCH 4 SDCCH/4. (*From:* [4].)

F: TDMA frame for frequency correction burst
B: TDMA frame for BCCH
D: TDMA frame for SDCCH
R: TDMA frame for RACH

S: TDMA frame for synchronization burst
C: TDMA frame for CCCH
A: TDMA frame for SACCH/C

1.5.3.6 Summary of Logical Channel Combinations

The different allowed combinations of logical channels on a physical channel are as follows:

- TCH/F + FACCH/F + SACCH/TF;
- TCH/H+ FACCH/H + SACCH/TH;
- FCCH + SCH + BCCH + CCCH;
- FCCH + SCH + BCCH + CCCH + SDCCH + SACCH;
- BCCH + CCCH;
- SDCCH+ SACCH.

Other logical channels exist for packet-switched services (GPRS, EDGE, or DTM), and will be described later.

1.5.4 Voice Digital Communication Chain

The functions performed by the physical layer during transmission are presented in this section, and the TCH/FS voice channel will be used for the sake of example. The steps of the communication chain, as outlined below, are presented in Figure 1.9.

1.5.4.1 Source Coding

First of all, speech must be digitized. On the basis of subjective speech quality and complexity issues, a *regular pulse excited–linear predictive coder* (RPE-LPC) with a long-term predictor loop was chosen. Basically, information from previous samples, which does not change very quickly, is used to predict the current sample. The coefficients of the linear combination of the previous samples, plus an encoded form of the residual (the difference between the predicted and actual sample), represent the signal. Speech is divided into 20-ms samples, each of which is encoded as 260 bits, giving a total bit rate of 13 Kbps. This is the so-called full-rate speech coding. Several other voice codecs are defined (half rate, enhanced full rate, adaptive multi-rate codecs), but these elements are beyond the scope of this book.

1.5.4.2 Channel Coding

In order to protect the voice against noise, interference, and multipath radio propagation conditions, the encoded speech transmitted over the radio

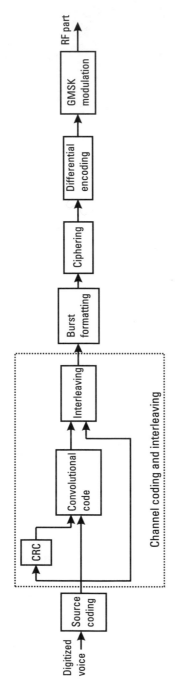

Figure 1.9 The voice transmission chain.

interface is protected from errors. Convolutional encoding and block inter-leaving are used to achieve this protection. As discussed above, the speech codec produces a 260-bit block for every 20-ms speech sample. From sub-jective testing, it was found that some bits of this block were more impor-tant for perceived speech quality than others. The bits are therefore divided into three classes:

- Class Ia: 50 bits, most sensitive to bit errors;
- Class Ib: 132 bits, moderately sensitive to bit errors;
- Class II: 78 bits, least sensitive to bit errors.

Class Ia bits have a 3-bit cyclic redundancy code added for error detec-tion. If an error is detected, the frame is determined to be incomprehensible, and it is discarded. In such a case it is replaced by a slightly attenuated ver-sion of the previous correctly received frame.

These 53 bits, together with the 132 class Ib bits and a 4-bit tail sequence (a total of 189 bits), are input into a 1/2-rate convolutional encoder of constraint length 4. Each input bit is encoded as 2 output bits, based on a combination of the previous 4 input bits. The convolutional encoder thus outputs 378 bits. The 78 remaining class II bits, which are unprotected, are added to these 378 bits. A total of 456 encoded bits are therefore produced for every 20ms speech sample. This represents a bit rate of 22.8 Kbps. In order to protect against burst errors common to the radio interface, each block of 456 bits is interleaved.

1.5.4.3 Interleaving

The 456 coded bits are permutated, and divided into eight blocks of 57 bits, and these blocks are transmitted in eight consecutive bursts, as described in Figure 1.10. Since each burst can carry two 57-bit blocks, each burst carries traffic from two different speech frames. The benefit of doing this is that it provides time diversity. If a sequence of several consecutive bits is corrupted by the degraded propagation conditions (fading, for example) during a given period of time, interleaving ensures that the errors will be randomly distrib-uted over the block of 456 bits. This property is required for a better perfor-mance of the decoding algorithm. The decoding of convolutional codes is indeed often performed by the Viterbi algorithm (refer to the case study on this subject in Chapter 4), for which the errors occur in packets. Thus, a ran-dom distribution of the corrupted bits over a block at the input of the

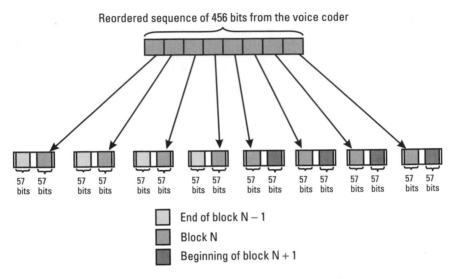

Reordered sequence of 456 bits from the voice coder

End of block N − 1
Block N
Beginning of block N + 1

Figure 1.10 Interleaving scheme on the TCH.

decoder leads to an increased decoding efficiency, and so to better performance in terms of bit error rate.

1.5.4.4 Burst Formatting

We have seen that the coded bits are interleaved and transmitted over bursts. The format of the burst-carrying TCH traffic, or *normal burst* (NB), is shown in Figure 1.11. NBs are used on most of the logical channels, except the RACH, SCH, and FCCH. The NB is constituted of two data blocks of 57 bits, carrying the coded voice samples, as described above. The middle of the burst contains the training sequence of 26 bits, known by the receiver, used to estimate the distortion introduced by the radio channel and for time synchronization. At the beginning and at the end of the burst, two sequences of 3 bits, known as the tail bits, all equal to 0, are used. Two bits, one before and one after the training sequence, form the *stealing flags* (SF). In Section 1.5.3.3 we saw that those bits allow the FACCH and TCH to be multiplexed, with the frame-stealing concept. When a speech frame is stolen for the transmission of a FACCH block, this is signaled by the SF bits. When one stealing bit is equal to one, this means that half of the burst is used for FACCH; otherwise it is used for TCH. An example of the multiplexing of FACCH and TCH is shown in Figure 1.12. Note that a FACCH is transported over eight half bursts. The burst duration is 148 bits, or 546.46 μs,

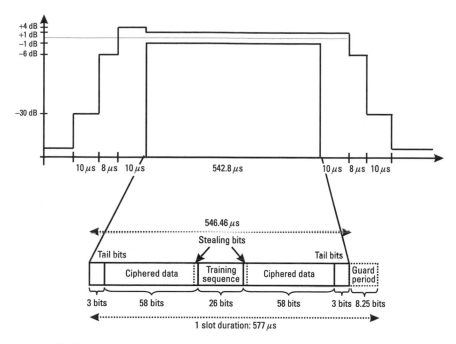

Figure 1.11 Structure of an NB.

and it is followed by a guard period of 8.25 bits, to allow for the burst ramping up and down between the time slots, as represented in Figure 1.11.

1.5.4.5 Ciphering

For security reasons, ciphering is used to modify the data parts of the burst, with a binary addition between a pseudorandom bit sequence and the 114 data bits. The same operation is performed by the receiver for deciphering. The pseudorandom bit sequence is different in the uplink and in the downlink.

1.5.4.6 Differential Encoding

All the bits d_i of the burst are differentially encoded. The output of the differential encoder is

$$\hat{d}_i = d_i \oplus d_{i-1} \quad \left(d_i \in \{0,1\} \right) \tag{1.1}$$

where $\oplus$ denotes modulo 2 addition.

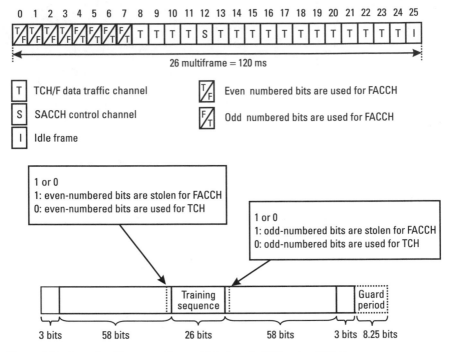

Figure 1.12 Example of frame stealing: multiplexing of a TCH and an FACCH.

The result is mapped onto +1 and −1 values, to form the modulating data value α_i input to the modulator, as follows:

$$a_i = 1 - 2\hat{d}_i \quad \left(a_i \in \{-1, +1\}\right) \tag{1.2}$$

1.5.4.7 GMSK Modulation

The digital signal is modulated onto the analog carrier frequency using *Gaussian-filtered minimum shift keying* (GMSK), with a symbol period of 48/13 μs (i.e., 270.8333 kHz). This modulation was selected over other modulation schemes as a compromise between spectral efficiency, complexity of the transmitter, power consumption for the MS, and limited out of channels emissions. These radio emissions, outside of the allocated channel, must be strictly controlled so as to limit adjacent channel interference and allow for the coexistence of GSM and the other systems. The spectrum due to the modulation mask requirement is presented in Figure 1.13, along with an ideal GMSK spectrum.

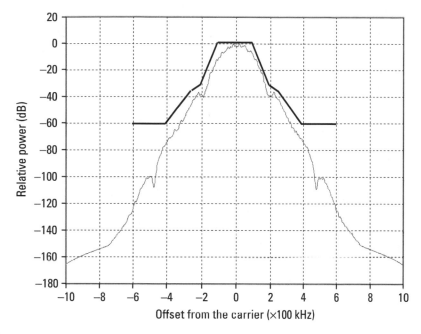

Figure 1.13 Ideal GMSK spectrum and required spectrum mask.

On the receive section, the opposite operations are performed, namely, demodulation, burst deformatting, de-interleaving, channel decoding, and source decoding.

1.5.4.8   Demodulation

Because of the various obstacles in the environment, many reflected signals, each with a different time delay and phase, arrives at the receiver. This is called multipath, and it is time variant, since the terminal is mobile, by definition. To cope with these time-varying propagation conditions, the receiver uses an equalizer. An equalizer is an algorithm that uses the receive sampled symbols to estimate the sequence of bits that was transmitted by the peer entity, by suppressing the *intersymbol interference* (ISI). Equalization is therefore used to extract the desired signal from the unwanted reflections. To do this, the receiver uses a known sequence of transmitted bits, the training sequence, to estimate the *channel impulse response* (CIR). This known signal is the 26-bit training sequence transmitted in the middle of every NB. The CIR is then used by the equalizer to retrieve the transmitted symbols. The estimation of the CIR also allows the finest time synchronization of the mobile to the BTS (the mobile can detect and correct a delay of several symbol periods). Note that the actual implementation of the equalizer is not specified in the GSM system, and it is up to the mobile phone or BTS vendor to implement a solution that will achieve the specified receiver performance (see Section 1.5.6.2).

1.5.5   Bursts Format

There are four burst formats the purposes of which are defined as follows:

1. The NB is used to carry information on traffic and control channels, except for RACH, SCH, and FCCH. It contains 114 encrypted symbols and includes a guard time of 8.25 symbol duration ($\approx$30,46 µs). A training sequence of 26 symbols is present in the middle of the burst (see Figure 1.11, Section 1.5.4).

2. The *frequency correction burst* (FB), as shown in Figure 1.14(a), contains a sequence of 142 fixed bits. This sequence is made of alternating ones and zeros (1, 0, 1, 0, ... 1, 0), so that after the differential encoding and GMSK modulation, the RF signal is equivalent to an unmodulated carrier, shifted by 67.7 kHz above the carrier frequency. This characteristic is used to help the mobile synchronize in frequency with the BTS, as explained in Section 1.5.7. The FB is transmitted over the FCCH.

3. The *synchronization burst* (SB) is needed for time synchronization of the mobile on the SCH. It contains a long training sequence and carries the information of the TDMA FN and *base station identity code* (BSIC), as can be seen in Figure 1.14(b).

4. The *access burst* (AB), presented in Figure 1.14(c), is used for *random access* (or the RACH) and is characterized by a longer guard period (68.25 bit duration or 252 µs), allowing the estimation of the *timing advance* (TA) by the BTS (see Section 1.5.6.3).

In Figure 1.14, we see a guard period at the end of a burst. During this period, the transmission is attenuated in several steps, as specified by the power-versus-time mask specification (see the example of NB, Figure 1.11).

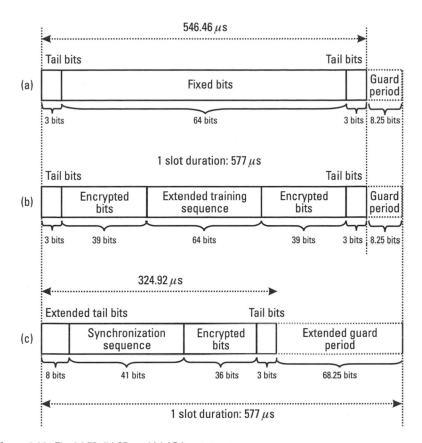

Figure 1.14 The (a) FB, (b) SB, and (c) AB burst structures.

1.5.6 RF Characteristics

1.5.6.1 Transmission Characteristics

Several classes of mobiles are defined, according to their maximum output power capability, as shown in Table 1.3. In GSM-900, most of the mobiles available on the market are class 4 handheld terminals, while class 2 terminals are used as vehicle-mounted equipment. The class 4 and 5 MSs are denoted as "small MS." In DCS-1800, the typical class is class 1.

These output power levels are maximum values, and can be reduced according to the commands that are sent by the network to the MSs. With

Table 1.3
MS Power Classes

Power Class	GSM-400, GSM-900, GSM-850 Nominal Maximum Output Power	DCS-1800 Nominal Maximum Output Power	PCS-1900 Nominal Maximum Output Power
1	—	1W (30 dBm)	1W (30 dBm)
2	8W (39 dBm)	0.25W (24 dBm)	0.25W (24 dBm)
3	5W (37 dBm)	4W (36 dBm)	2W (33 dBm)
4	2W (33 dBm)		
5	0.8W (29 dBm)		

these network commands, the MS operates at the lowest power level that maintains an acceptable signal quality.

These commands are based on the measurements that are performed by the MS and by the BTS. For instance, with a class 4 MS, the range of transmission can be several kilometers, but if the MS is getting closer to the BTS, it may receive a request from the network to decrease its output power level. This procedure, called power control, improves the performance of the system by reducing the interference caused to the other users. Moreover, it is a means of prolonging the battery life of the mobile. The power level can be stepped up or down in steps of 2 dB from the maximum power (depending on the MS class) down to a minimum of 5 dBm in GSM-400/900/850, and 0 dBm in DCS-1800/PCS-1900. The transmission of power control commands by the BTS is explained in Section 1.5.6.3.

For the BTS transceiver (TRX), the power classes are given in Table 1.4.

As an option, the BSS can utilize downlink RF power control, with up to 15 steps of power control levels with a step size of 2 dB. Note that this power control on the downlink is not used on the beacon frequency, which is always transmitted with constant output power.

Many other requirements on the transmit section are defined in the GSM specifications, such as the spectrum due to modulation constraint (see Figure 1.13), the modulation accuracy, the transmitter frequency error, and the spurious emissions requirements.

Table 1.4
TRX Power Classes

TRX Power Class	GSM-400, GSM-900, GSM- 850 Maximum Output Power	DCS-1800 and PCS-1900 Maximum Output Power
1	320 (< 640)W	20 (< 40)W
2	160 (< 320)W	10 (< 20)W
3	80 (< 160)W	5 (< 10)W
4	40 (< 80)W	2.5 (< 5)W
5	20 (< 40)W	
6	10 (< 20)W	
7	5 (< 10)W	
8	2.5 (< 5)W	

1.5.6.2   Reception Characteristics

Several types of propagation models have been defined, in order to measure the mobile and BTS performances. These models represent several environments:

- *Typical urban* (TUx);
- *Rural area* (RAx);
- *Hilly terrain* (HTx).

In the above definitions, the x stands for the velocity of the mobile, in km/h. The various propagation models are represented by a number of taps, each determined by their time delay and average power. The Rayleigh distributed amplitude of each tap varies according to a Doppler spectrum.

In addition to these multipath fading channels, the static channel was defined. This is a simple single-path constant channel. With this channel, the only perturbation comes from the receiver noise of the measured equipment.

One of the most important receiver performances that is specified is the sensitivity level, which determines the minimum level for which the receiver can demodulate a signal correctly. The sensitivity requirement, in GSM, is specified as an input level, in dBm, for which the measured equipment should reach a certain performance, in terms of bit error rate. For instance, for GSM400/900/850 power classes 4 or 5 mobiles and DCS-1800/PCS-1900

Table 1.5
Sensitivity-Level Performance Requirements

Logical Channel		Static	Propagation Conditions			
			TU (no FH)	TU50 (ideal FH)	RA250 (no FH)	HT100 (no FH)
FACCH/H	(FER)	0.1%	6.9%	6.9%	5.7%	10.0%
FACCH/F	(FER)	0.1%	8.0%	3.8%	3.4%	6.3%
SDCCH	(FER)	0.1%	13%	8%	8%	12%
RACH	(FER)	0.5%	13%	13%	12%	13%
SCH	(FER)	1%	16%	16%	15%	16%
TCH/F14.4	(BER)	10.0^{-5}	2.5%	2%	2%	5%
TCH/F9.6	(BER)	10.0^{-5}	0.5%	0.4%	0.1%	0.7%
TCH/FS	(FER)	0.1α%	6α%	3α%	2α%	7α%
Class Ib	(RBER)	$0.4/\alpha$%	$0.4/\alpha$%	$0.3/\alpha$%	$0.2/\alpha$%	$0.5/\alpha$%
Class II	(RBER)	2%	8%	8%	7%	9%

classes 1 or 2 mobiles, the sensitivity level is −102 dBm. For a normal BTS (that is not a micro- or a pico-BTS) the sensitivity level is −104 dBm, for all the frequency bands.

At these levels, different performances, according to both the logical channel and the propagation channel used for the measurement, must be met. Table 1.5 shows an example of performances that are reached at the sensitivity level, for GSM-900 and GSM-850. In this table, BER stands for bit error rate, FER for frame erasure ratio (i.e., incorrect-speech-frames ratio), and RBER for residual BER (defined as the ratio of the number of errors detected over the frames defined as "good" to the number of transmitted bits in the good frames). This table is an example; similar tables exist for the other logical channels and for the different frequency bands. Note that frequency hopping may be used for the sensitivity performance measurements.

Note that in this example, the parameter α is defined as $1 \leq \alpha \leq 1.6$ and allows a tradeoff between the number of erased speech frames (i.e., decoded as wrong, and therefore not transmitted to the voice decoder) and the quality of the nonerased frames.

Another important characteristic of the receiver concerns its performance in the presence of an interferer. This is specified either for a cochannel

Table 1.6
Interference Performance Requirements

		Propagation Conditions				
Type of Channel		**TU3 (No FH)**	**TU3 (Ideal FH)**	**TU50 (No FH)**	**TU50 (Ideal FH)**	**RA250 (No FH)**
FACCH/H	(FER)	22%	6.7%	6.7%	6.7%	5.7%
FACCH/F	(FER)	22%	3.4%	9.5%	3.4%	3.5%
SDCCH	(FER)	22%	9%	13%	9%	8%
RACH	(FER)	15%	15%	16%	16%	13%
SCH	(FER)	17%	17%	17%	17%	18%
TCH/F 14.4	BER)	10%	3%	4.5%	3%	3%
TCH/F 9.6	(BER)	8%	0.3%	0.8%	0.3%	0.2%
TCH/FS	(FER)	$21\alpha\%$	$3\alpha\%$	$6\alpha\%$	$3\alpha\%$	$3\alpha\%$
Class Ib	(RBER)	$2/\alpha\%$	$0.2/\alpha\%$	$0.4/\alpha\%$	$0.2/\alpha\%$	$0.2/\alpha\%$
Class II	(RBER)	4%	8%	8%	8%	8%

interference (i.e., an interference situated at the same frequency as the signal of interest) or an adjacent channel interference (situated at 200 or 400 kHz from the carrier of interest). The level of the useful signal is set 20 dB higher than for the sensitivity evaluation, and a GMSK interfering signal is added, either at the same frequency or with an offset of 200 or 400 kHz from the carrier. For the cochannel test, the carrier to interference ratio C/Ic is set to 9 dB. Under these conditions, the performance of Table 1.6 must be met. Again, this table does not contain all the logical channels, and concerns the GSM-900 and GSM-850 only. Similar performance requirements are defined for the other cases.

This table is also applied in the case of an adjacent channel interference. In this case, the C/I is set to −9 dB if the interferer is 200 kHz from the carrier, and −41 dB if it is 400 kHz from the carrier.

1.5.6.3 Control of the Radio Link

This section describes some of the procedures that are in use to improve the efficiency of the system, by adapting the transmission between the mobile and the BTS to the continuously varying radio environment.

Compensation for the Propagation Delay

Due to the distance between the MS and the BTS, there is a propagation delay that is equal to d/c seconds, where d is the MS to BTS distance in meters, and c is the speed of light ($c = 3 \cdot 10^8$ m.s^{-1}). Without any compensation of this delay, the bursts transmitted by two different MSs, in the same TDMA frame on two consecutive slots, could interfere with one another.

Let us take the example of one MS situated 25 km away from the BTS, transmitting on time slot 0 of a given channel frequency. Another MS is located, say, 1 km away from the BTS, and transmitting on time slot 1 of that same frequency. The second MS transmission will experience a very short delay (around 3.33 μs), but the burst on time slot 0, from MS 1, will be received by the BTS 83.33 μs after it has been transmitted. This means that at the BTS receiver, the burst on time slot 0 will interfere with the beginning of the burst of time slot 1, for a period of about 80 μs. This example is represented in Figure 1.15.

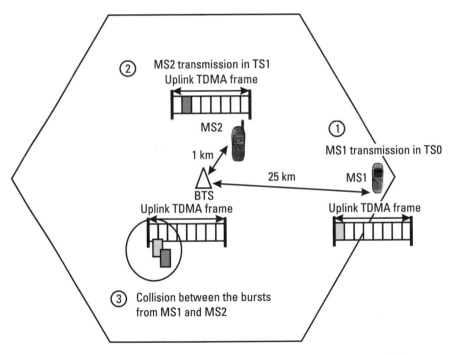

Figure 1.15 Propagation delay difference between two MSs transmitting to the same BTS.

In order to cope with this problem, the network manages a parameter for each mobile called the TA. This parameter represents the transmission delay between the BTS and the MS, added to the delay for the return link.

The estimation of the delay is performed by the BTS upon reception of an AB on the RACH. As described in Section 1.5.5, this burst is characterized by a longer guard period (68.25-bit duration or 252 µs) to allow burst transmission from a mobile that does not know the TA at the first access. The received AB allows the BTS to estimate the delay by means of a correlation with the training sequence.

The TA value, between 0 and 63 symbol periods (i.e., between 0 and 232.615 µs by steps of 48/13 µs), is transmitted on the AGCH. It allows the MS to advance its time base, so that the burst received at the BTS arrives exactly three time slots after the BTS transmit burst, as shown in Figure 1.16. A distance of 35 km between the MS and the BTS is therefore possible. (The 232.675 µs allows to compensatefor a distance of around 70 km, including the forward and return links.)

After this first propagation delay estimation, the BTS continuously monitors the delay of the NBs sent by from the MS on the other logical

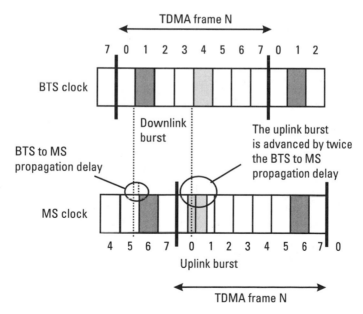

Figure 1.16 Correction of MS transmission timing to compensate for propagation delay.

channels. If the delay changes by more than one symbol period, a new value of the TA is signaled to the MS on the SACCH.

MS Power Control

As explained in Section 1.5.6.1, the MS can vary its transmit output power from a maximum defined by its power class, by steps of 2 dB. During a communication, the MS and BTS measure the received signal strength and quality (based on the bit error ratio) and pass the information to the BSC, which ultimately decides if and when the power level should be changed. A command is then sent to the MS on the SACCH.

Power control is a difficult mechanism to implement, since there is a possibility of instability. This arises from having MS in cochannel cells, alternatively increasing their power in response to increased cochannel interference. Suppose that mobile A increases its power because the corresponding BTS receives a cochannel interference caused by mobile B, in another cell. Then the BTS receiving the signal from mobile B might request mobile B to increase its power, and so forth. This is the reason why some coordination is required at the BSC level.

Note that for an access request on the RACH, the MS uses the maximum power level defined by the parameter MS_TXPWR_MAX_CCH broadcast by the network.

Frequency Hopping

The radio environment depends on the radio frequency. In order to avoid important differences in the quality of the channels, a feature called slow *frequency hopping* (FH) was introduced. The slow FH changes the frequency with every TDMA frame, which also has the effect of reducing the cochannel interference. This capability is optionally used by the operator, and is not necessarily implemented in all the cells of the network, but it must be supported by all the MSs. The main advantage of FH is to provide diversity on one transmission link (especially to increase the efficiency of coding and interleaving for slowly moving MSs) and also to average the quality on all the communications through interference diversity.

The principle of slow FH is that every mobile transmits its time slots according to a sequence of frequencies that it derives from an algorithm. The FH sequences are orthogonal inside one cell (i.e., no collisions occur between communications of the same cell) and independent from one cell to a cochannel cell (i.e., a cell using the same set of RF channels or cell allocation). The hopping sequence is derived by the mobile from parameters broadcast at the channel assignment, namely, the mobile allocation (set of N frequencies on which to hop), the *hopping sequence number* (HSN) of the cell (which allows different sequences on cochannel cells), and the index offset (to distinguish the different mobiles of the cell using the same mobile allocation) or *mobile allocation index offset* (MAIO). Based on these parameters and on the FN, the MS knows which frequency to hop in each TDMA frame.

It must be noted that the basic physical channel supporting the BCCH does not hop.

1.5.7 MS Cell Synchronization Procedure

In synchronizing to a cell, the MS first searches for the FB on the FACCH. This allows a first timing synchronization, but most of all, it allows the mobile to adjust its oscillator to be synchronized in the frequency domain with the BTS. This is possible because, as described in Section 1.5.6, the fixed sequence of the FB has been chosen so that the modulating bit sequence at the GMSK modulator input is constant. This results in a continuous $\pi/2$ phase rotation, which in the frequency domain is equivalent to an unmodulated carrier with a +1 625/24 kHz frequency offset, above the nominal carrier frequency. Once the MS has identified the FB, it uses this property to estimate its frequency drift with regard to the BTS.

In the TDMA frame immediately following the occurrence of an FB, on the same carrier frequency, the SB is transmitted over the air interface, on the SCH. This burst is identified by the MS with its extended 64-bit training sequence code. The MS received samples of the SB, correlated with the known training sequence, allow for the timing of the mobile to be adjusted to the base station with good precision. At this point, the MS and the BTS are synchronized in the time domain, except that the propagation delay between them is not compensated. This is performed with the TA scheme, as discussed in Section 1.5.6.3, when the MS sends an AB on the RACH.

The decoding of the SB enables a logical synchronization of the mobile to the cell, since it gives the elements to estimate the TDMA FN (see Section 1.5.3.1), which allows the MS to determine the position of the SCH in the hyperframe. The SCH also contains the *BTS identity code* (BSIC): these 6 bits (before channel coding) consist of the PLMN color code with range 0 to 7 and of the BS color code with range 0 to 7 (3 bits each).

1.5.8   Summary of MS Operations in Idle Mode

In idle mode, as opposed to dedicated mode, the mobile has no channel of its own. It is required to

- Synchronize in time and frequency to a given cell, selected as the best suitable cell with regard to a set of criteria (based on the beacon received power at the MS). This is termed "camping onto" a cell. This process of evaluating different cells and choosing the best suitable one is called cell selection, or reselection if it is performed again, due to the degradation of the link quality with the previously selected cell. The MS during idle mode continuously measures the radio link quality of the serving cell and the surrounding cells, so that cell reselection criteria are evaluated periodically.

- Listen to possible incoming calls from the network. The notification of an incoming call is usually referred to as paging.

1.5.8.1   Selection of the PLMN

When the mobile is switched on, the first operation that it performs is identification or selection of a PLMN. Most of the time, the PLMN will be the home PLMN (i.e., the network to which the user has subscribed). In such a case, no selection is needed, as information about the network is stored in the SIM card. If it is not the case, because the user is traveling in a different area, the MS will scan all the frequencies in order to detect the surrounding beacon channels (detection of FB and SB, as described in Section 1.5.7). The MS is then able to decode the PLMN identifiers, and either choose the first PLMN in the priority ordered list of the SIM card, or ask the user which PLMN is preferred among all the detected PLMNs. The selection is then stored, in order to be used at the next terminal switch on. In any case, the user can explicitly ask for a given PLMN selection.

1.5.8.2   Principles of Cell Selection and Reselection

Once the PLMN is selected, the MS must select a cell. Two scenarios are possible:

1. The beacon channel frequencies are stored in the MS, because it has already performed a selection in the previous terminal activity. In this case, the MS will perform measurement on these frequencies, to determine which cell is the most suitable with regard to certain cri-

teria. Once the best cell is selected, the MS performs registration and "camps on" this cell. Note that if the stored frequency list of beacon carriers is not detected by the MS, it will perform a PLMN selection as described above.

2. It is the first time the PLMN is accessed. The carriers of the system are all scanned, in order to detect the beacon channels, and the received signal strength of these channels is added in an ordered list. Once this is achieved, the cell selection can be performed, as in the previous case. In order to speed up the process, a list of the RF channels containing BCCH carriers of the same PLMN is broadcast in the system information messages.

When an MS is camping on a cell, it can receive paging blocks on the PCH, or initiate call setup for outgoing calls by sending an AB on the RACH. It still regularly monitors the signal level on the surrounding beacon carriers, and evaluates the reselection criteria. The reselection is triggered if one of the following events occurs:

- The path loss criterion parameter C1 indicates that the path loss to the cell has become too high.
- There is a downlink signaling failure (i.e., the success rate of the MS in decoding signaling blocks drops too low).
- The cell camped on has become barred (this means that the operator decides not to allow MSs to camp on this cell).
- There is a better cell, in terms of the path loss criterion C2 in the same registration area.
- A random access attempt is still unsuccessful after a given number of repetitions, specified by a broadcast parameter.

The criteria for cell selection and reselection (path loss criterion C1 and reselection criterion C2) are based on the measurements performed by the MS on the BCCH frequency. (As stated earlier, the beacon frequency is transmitted with its maximum output power by the BTS.) Details on these criteria for cell selection are given in Chapter 5.

1.5.8.3 Monitoring of Paging Blocks

We discussed the mapping of the CCCH on the 51 multiframe in Section 1.5.3.5, and in particular the case of the PCH. This logical channel is used to convey paging blocks on the downlink. These blocks are used to notify the MS of an incoming call. In order to conserve the MS's power, a PCH is divided into subchannels, each corresponding to a group of MSs. Each MS will then only "listen" to its subchannel and will stay in the sleep mode during the other subchannels of the PCH. This is called the *discontinuous reception* (DRX) mode. The mobile knows in which group it belongs by determining the parameter CCCH_GROUP. It is estimated with an algorithm, which inputs are the mobile IMSI and the parameter BS_CC_CHANS, broadcast on the BCCH. This parameter defines the number of basic physical channels supporting CCCH.

Mobiles in a specific CCCH_GROUP will listen for paging messages and make random accesses only on the specific CCCH to which the CCCH_GROUP belongs. This algorithm is detailed in Chapter 4. Note that the MS is not authorized to use the DRX mode of operation while performing the cell-selection algorithm.

1.5.9 Measurements Performed by MS During Communication

When assigned a TCH or SDCCH, during the time slots that are not used for these channels and for the associated SACCH, the MS performs measurements on all the adjacent BCCH frequencies. These measurements are then sent to the network by means of the SACCH, and are interpreted by the NSS for the power control and handover procedures. Measurements are performed in each TDMA frame, (see Figure 1.17) and are referred to as monitoring, which consists of estimating the receive signal strength on a given frequency. The list of frequencies to be monitored is broadcast on the BCCH, by means of the BCCH allocation (BA) list, which contains up to 32 frequencies. The frequencies are monitored one after the other, and the measured samples are averaged prior to the reporting to the network, on an uplink SACCH block, under form of a value called RXLEV. The MS therefore measures the received signal level from surrounding cells by tuning and listening to their BCCH carriers. This can be achieved without interbase station synchronization. The measurements are reported at every reporting period.

For a TCH/FS, the reporting period duration is 104 TDMA frames (480 ms).

It is essential that the MS identify which surrounding BSS is being measured in order to ensure reliable handover. Because of frequency reuse

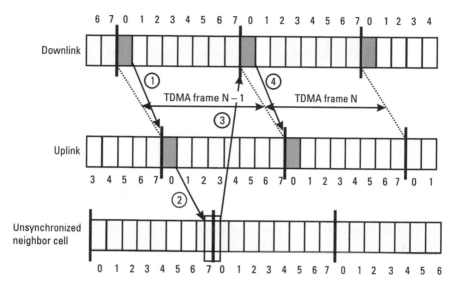

Figure 1.17 Monitoring during a TDMA frame.

with small cluster sizes, the BCCH carrier frequency may not be sufficient to uniquely identify a surrounding cell. The cell in which the MS is situated may have more than one surrounding cell using the same BCCH frequency. It is therefore necessary for the MS to synchronize to (using the method explained in Section 1.5.7) and demodulate surrounding BCCH carriers to identify the BSIC in the SB. In order to do so, the MS uses the idle frames. These frames are termed "search" frames. Note that a window of nine consecutive slots is needed to find time slot 0 on the BCCH frequency (remember that time slot 0 carries the SCH and FCCH), since the beacon channels are not necessarily synchronized with one another. One important characteristic to notice is that the SCH and FCCH are mapped onto the 51 multiframe, and that the idle frame of the mobile during communication is occurs in on the 26 multiframe. Since 26 and 51 are mutually prime numbers, this means a search frame will be available every 26 modulo 51 frame on the beacon channel.

For instance, let us imagine that an idle frame occurs in the frame 0 of the 51 multiframe. The next idle frames will be programmed on frames 26, 1, 27, 2, and so on. Therefore, after a certain number of search frames, the MS will necessarily decode an FB and an SB.

Another measured parameter during a TCH or SDCCH is the RXQUAL, which represents an indication of the quality of the received link, in terms of BER. For each channel, the measured received signal quality is averaged on that channel over the reporting period of length one SACCH multiframe defined above.

References

[1]   Mouly, M., and M. B. Pautet, *The Global System for Mobile Communications*, 1993.

[2]   Lagrange, X., P. Godlewski, and S. Tabbane, *Réseaux GSM-DCS*, 5th ed., Hermès, 2001 (available in French only).

[3]   3GPP TS 05.05 Radio Transmission and Reception (R99).

[4]   3GPP TS 05.01 Physical Layer on the Radio Path; General Description (R99).

2

GPRS Services

2.1 Use of GPRS

The GPRS provides a set of GSM services for data transmission in packet mode within a PLMN. In packet-switched mode, no permanent connection is established between the mobile and the external network during data transfer. Instead, in circuit-switched mode, a connection is established during the transfer duration between the calling entity and the called entity. In packet-switched mode, data is transferred in data blocks, called packets. When the transmission of packets is needed, a channel is allocated, but it is released immediately after. This method increases the network capacity. Indeed, several users can share a given channel, since it is not allocated to a single user during an entire call period.

One of the main purposes of GPRS is to facilitate the interconnection between a mobile and the other packet-switched networks, which opens the doors to the world of the Internet. With the introduction of packet mode, mobile telephony and Internet converge to become mobile Internet technology. This technology introduced in mobile phones allows users to have access to new value-added services, including:

- *Client-server services,* which enable access to data stored in databases. The most famous example of this is access to the *World Wide Web* (WWW) through a browser.
- Messaging services, intended for user-to-user communication between individual users via storage servers for message handling.

Multimedia Messaging Service (MMS) is an example of a well-known messaging application.

- Real-time conversational services, which provide bidirectional communication in real-time. A number of Internet and multimedia applications require this scheme such as voice over IP and video conferencing.

- Tele-action services, which are characterized by short transactions and are required for services such as SMS, electronic monitoring, surveillance systems, and lottery transactions.

GPRS allows for radio resource optimization by using packet switching for data applications that may present the following transmission characteristics:

- Infrequent data transmission, as when the time between two transmissions exceeds the average transfer delay (e.g., messaging services);

- Frequent transmission of small data blocks, in processes of several transactions of less than 500 octets per minute (e.g., downloading of several HTML pages from a browsing application);

- Infrequent transmission of larger data blocks, in processes of several transactions per hour (e.g., access of information stored in database centers);

- Asymmetrical throughput between uplink and downlink, such as for data retrieval in a server where the uplink is used to send signaling commands and the downlink is used to receive data as a response of the request (e.g., WEB/WAP browser).

As the GPRS operator optimizes radio resources by sharing them between several users, he is able to propose more attractive fees for data transmission in GPRS mode than in circuit-switched mode. Indeed, the invoicing in circuit-switched mode takes into account the connection time between the calling user and the called user. Studies on data transmission show that data are exchanged from end to end during 20% of a circuit-switched connection time. For example, a user browses the WWW, downloads an HTML page identified by a *uniform resource locator* (URL), reads the content of the HTML page, then downloads a new HTML page to read. In this example no data is exchanged from end to end between the two

HTML page downloads. For this type of application, a more appropriate invoicing would take into account the volume of data exchanged instead of the circuit-switched connection time. In packet mode, the GPRS user may be invoiced according to the requested service type, the volume of data exchanged.

2.2   GPRS MS Classes

Three GPRS classes have been defined: class A, class B, and class C.

The class A mobile can support simultaneously a communication in circuit-switched mode and another one in packet-switched mode. It is also capable of detecting in idle mode an incoming call in circuit or packet-switched mode.

The class B mobile can detect an incoming call in circuit-switched mode or in packet-switched mode during the idle mode but cannot support them simultaneously. The circuit and packet calls are performed sequentially. In some configurations desired by the user, a GPRS communication may be suspended in order to perform a communication in circuit-switched mode and then may be resumed after the communication release in circuit-switched mode.

The class C mobile supports either a communication in circuit-switched mode or in packet-switched mode but is not capable of simultaneously supporting communications in both modes. It is not capable of simultaneously detecting the incoming calls in circuit-switched and packet-switched mode during idle mode. Thus a class C mobile is configured either in circuit-switched mode or in packet-switched mode. The mode configuration is selected either manually by the user or automatically by an application.

A mobile defined in class A or class B is IMSI attached for GPRS services, and non-GPRS services while a mobile defined in class C is IMSI attached if it operates in circuit-switched mode or IMSI attached for GPRS services if it operates in packet-switched mode. (*Note:* An MS that is IMSI attached means that it is attached to the GSM network.)

2.3   Client-Server Relation

The GPRS packet-transmission mode relies on the "client/server" principle from the computer world, rather than the "calling/called" principle in use in

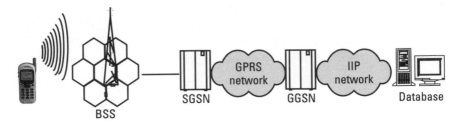

Figure 2.1   GPRS mobile configured as a client.

the telephony domain. The client sends a request to the server, which processes the request and sends the result to the client. Thus the mobile may be configured according to the application either in client mode or in server mode.

The mobile may be configured in client mode to have access to the Internet or an intranet or database by initiating a GPRS communication. Usually, the GPRS mobile is configured as a client. Figure 2.1 shows a GPRS MS configured in client mode.

The mobile may also be configured in server mode for vertical application to telemetry monitoring. In this type of application, the mobile may be connected to different pieces of equipment, such as a camera for monitoring or a captor for measurements. The mobile may configure a piece of equipment in order to process the request and then send back the result to the client. In order to interpret a request from a client, the mobile must be able to route information from the network toward the recipient application. In server mode, the MS must be IMSI attached for GPRS services in order to receive the requests from a client.

2.4   Quality of Service

The network associates a certain *quality of service* (QoS) with each data transmission in GPRS packet mode. The appropriate QoS is characterized according to a number of attributes negotiated between the MS and the network. Figure 2.2 characterizes the application in terms of error tolerance and delay requirements.

A first list of attributes is defined in Release 97/98 of the 3GPP recommendations. It was replaced in the release 99 by new attributes.

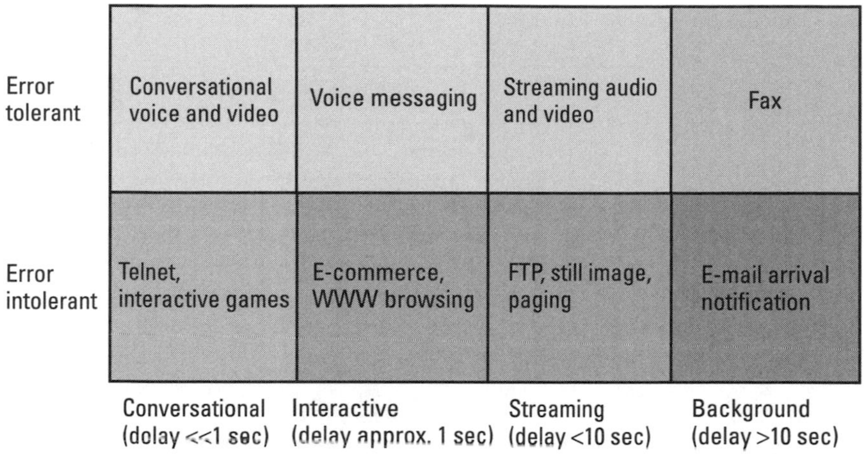

Error tolerant	Conversational voice and video	Voice messaging	Streaming audio and video	Fax
Error intolerant	Telnet, interactive games	E-commerce, WWW browsing	FTP, still image, paging	E-mail arrival notification
	Conversational (delay <<1 sec)	Interactive (delay approx. 1 sec)	Streaming (delay <10 sec)	Background (delay >10 sec)

Figure 2.2 Applications in terms of QoS requirements. (*From:* [1].)

2.4.1 Attributes in Release 97/98

In Release 97/98 of the 3GPP recommendations, QoS is defined according to the following attributes:

- *Precedence class.* This indicates the packet transfer priority under abnormal conditions, as for example during a network congestion load.

- *Reliability class.* This indicates the transmission characteristics; it defines the probability of data loss, data delivered out of sequence, duplicate data delivery, and corrupted data. This parameter enables the configuration of layer 2 protocols in acknowledged or unacknowledged modes.

- *Peak throughput class.* This indicates the expected maximum data transfer rate across the network for a specific access to an external packet switching network (from 8 to 2,048 Kbps).

- *Mean throughput class.* This indicates the average data transfer rate across the network during the remaining lifetime of a specific access to an external packet switching network (best effort, from 0.22 bps to 111 Kbps).

Table 2.1
Delay Classes

Delay Class	Delay (Maximum Values)			
	SDU size: 128 octets		SDU size: 1,024 octets	
	Mean Transfer Delay (s)	95 Percentile Delay (s)	Mean Transfer Delay (s)	95 Percentile Delay (s)
1. (Predictive)	< 0.5	< 1.5	< 2	< 7
2. (Predictive)	< 5	< 25	< 15	< 75
3. (Predictive)	< 50	< 250	< 75	< 375
4. (Best Effort)	Unspecified			

From: [2].

- *Delay class.* This defines the end-to-end transfer delay for the transmission of *service data units* (SDUs) through the GPRS network. The SDU represents the data unit accepted by the upper layer of GPRS and conveyed through the GPRS network. Table 2.1 shows the delay classes.

The delay class for data transfer gives some information about the number of resources that have to be allocated for a given service. Predictive value in delay class means that the network is able to ensure an end-to-end delay time for the transmission of SDUs; best effort means that the network is not able to ensure a value for an end-to-end transfer delay; in this case transmission of SDUs depends on network load.

2.4.2   Attributes in Release 99

The attributes of GPRS QoS were modified in Release 99 of the 3GPP recommendations in order to be identical to the ones defined for UMTS. The attributes described below apply to both GPRS and UMTS standards. Table 2.2 gives the characteristics of the different classes.

Four classes of traffic have been defined for QoS:

1. *Conversational class.* These services are dedicated to bidirectional communication in real time (e.g., voice over IP and videoconferencing).

Table 2.2

Traffic Classes

Traffic Class	Real-Time Conversational	Real-Time Streaming	Interactive Best Effort	Background Best Effort
Fundamental Characteristics	No transfer delay variation between the sender and the receiver; stringent and low delay transfer	No transfer delay variation between the sender and the receiver	Request response pattern; preserve pattern content	No time constraint; preserve pattern content
Example of Applications	Conversational voice and video-phone	One-way video, audio streaming, still image, and bulk data	Web browsing, voice messaging and dictation, server access, and e-commerce	E-mail, SMS, and fax

2. *Streaming class.* These services are dedicated to unidirectional data transfer in real time (e.g., audio streaming, one-way video).

3. *Interactive class.* These services are dedicated to the transport of human or machine interaction with remote equipment (e.g., Web browsing, access to a server, access to a database).

4. *Background class.* These services are dedicated to machine-to-machine communication that is not delay sensitive (e.g., e-mail and SMS).

Table 2.3 lists the expected performance for conversational services.
Table 2.4 lists the expected performance for streaming services.
Table 2.5 lists the expected performance for interactive services.
The Release 99 of 3GPP recommendations defines attributes for QoS such as traffic class, delivery order, SDU format information, SDU error ratio, maximum SDU size, maximum bit rate for uplink, maximum bit rate for downlink, residual bit error ratio, transfer delay, traffic-handling priority, allocation/retention priority, and guaranteed bit rate for uplink and guaranteed bit rate for downlink.

- *Traffic class* indicates the application type (conversational, streaming, interactive, background).
- *Delivery order* indicates if there is in-sequence SDU delivery or not.

Table 2.3
End User Performance Expectations—Conversational/Real-Time Services

Medium	Application	Degree of Symmetry	Data Rate	Key Performance Parameters and Target Values		
				End-to-End One-Way Delay	Delay Variation Within a Call	Information Loss
Audio	Conversa-tional voice	Two-way	4–25 Kbps	<150 ms preferred <400 ms limit Note 1	< 1 ms	< 3% of frame error rate
Video	Videophone	Two-way	32–384 Kbps	< 150 ms preferred < 400 ms limit Lip-synch: <100 ms		< 1% of frame error rate
Data	Telemetry—two-way control	Two-way	<28.8 Kbps	< 250 ms	N/A	Zero
Data	Interactive games	Two-way	< 1 KB	< 250 ms	N/A	Zero
Data	Telnet	Two-way (asymmetric)	< 1 KB	< 250 ms	N/A	Zero

From: [1].

- *Delivery of erroneous SDUs* indicates if erroneous SDUs are delivered or discarded.

- *SDU format information* indicates the possible exact sizes of SDUs.

- *SDU error ratio* indicates the maximum allowed fraction of SDUs lost or detected as erroneous.

- *Maximum SDU size* indicates the maximum allowed SDU size (from 10 octets to 1,520 octets).

- *Maximum bit rate for uplink* indicates the maximum number of bits delivered to the network within a period of time (from 0 to 8,640 Kbps).

Table 2.4
End User Performance Expectations—Streaming Services

Medium	Application	Degree of Symmetry	Data Rate	Key Performance Parameters and Target Values		
				One-Way Delay	Delay Variation	Information Loss
Audio	High-quality streaming audio	Primarily one-way	32–128 Kbps	< 10 s	< 1 ms	< 1% FER
Video	One-way	One-way	32–384 Kbps	< 10 s		< 1% FER
Data	Bulk data transfer/ retrieval	Primarily one-way		< 10 s	N/A	Zero
Data	Still image	One-way		< 10 s	N/A	Zero
Data	Telemetry— monitoring	One-way	<28.8 Kbps	< 10 s	N/A	Zero

From: [1].

- *Maximum bit rate for downlink* indicates the maximum number of bits delivered by the network within a period of time (from 0 to 8,640 Kbps).

- *Residual bit error ratio* indicates the undetected bit error ratio for each subflow in the delivered SDUs.

- *Transfer delay* indicates the maximum time of SDU transfer for 95th percentile of the distribution of delay for all delivered SDUs.

- *Traffic-handling priority* indicates the relative importance of all SDUs belonging to a specific GPRS bearer compared with all SDUs of other GPRS bearers.

- *Allocation/retention priority* indicates the relative importance of resource allocation and resource retention for the data flow related to a specific GPRS bearer compared with the data flows of other GPRS bearers (useful when resources are scarce).

- Guaranteed bit rate for uplink indicates the guaranteed number of bits delivered to the network within a period of time (from 0 to 8,640 Kbps).

Table 2.5
End User Performance Expectations—Interactive Services

Medium	Application	Degree of Symmetry	Data Rate	Key Performance Parameters and Target Values		
				One-Way Delay	Delay Variation	Information Loss
Audio	Voice messaging	Primarily no way	4–13 Kbps	< 1 sec for playback < 2 sec for record	< 1 ms	< 3% FER
Data	Web browsing—HTML	Primarily one-way		< 4 sec/page	N/A	Zero
Data	Transaction services—high priority (e.g., e-commerce and ATM)	Two-way		< 4 sec	N/A	Zero
Data	E-mail (server access)	Primarily one-way		< 4 sec	N/A	Zero

From: [1].

- Guaranteed bit rate for downlink indicates the guaranteed number of bits delivered to the network within a period of time (from 0 to 8,640 Kbps).

2.5   Third-Generation Partnership Project

The GPRS recommendations belong to the GSM recommendations. The maintenance of all GSM recommendations is now handled within the *Third Generation Partnership Project* (3GPP) organization, the partners of which are:

- ETSI, the European standardization entity;
- *Association of Radio Industries and Businesses* (ARIB) and *Telecommunication Technology Committee* (TTC), the Japanese standardization entities;

- *Telecommunication Technology Association* (TTA), the Korean standardization entity;

- T1, the American standardization entity;

- *China Wireless Telecommunication Standard* (CWTS) group, the Chinese standardization entity.

These standardization bodies have decided to collaborate within the 3GPP organization in order to produce specifications for a third-generation mobile system. At the beginning, the 3GPP organization was in charge of all specifications related to the third-generation mobile system for radio access technologies called Universal Terrestrial Radio Access (UTRA) and for evolution of GSM core networks. Since August 2000, specifications related to GSM radio access are also the responsibility of the 3GPP.

The 3GPP takes the place of the former Special Mobile Group (SMG) GSM organization. The 3GPP is organized around *technical specification groups* (TSGs) that deal with the following subjects:

- TSG SA (Service Architecture), dealing with service, architecture, security, and speech coding aspects;

- TSG RAN (Radio Access Network), focusing on UTRA radio access technologies;

- TSG CN (Core Network), dealing with core network specifications;

- TSG T (Terminal), covering applications, tests for 3G mobiles, and the USIM card;

- TSG GERAN (GSM EDGE Radio Access Network), focusing on GSM radio interface, A and Gb interfaces.

Thus GSM evolutions as GPRS are treated in all TSGs except TSG RAN, which deals exclusively with UTRAN access technologies such as *frequency division duplex* (FDD), *time-division duplex* (TDD), and CDMA 2000. The TSG GERAN deals exclusively with the GSM radio interface evolutions and with A and Gb interfaces.

The 3GPP recommendations are ranked according to a version reference. Each new version of 3GPP recommendations contains a list of new features or a list of improvements on existing features. Initially, the GSM recommendations versions were referenced in the following order: Phase 1; Phase 2; Release 96; Release 97; Release 98; Release 99. As the reference year for the new version of phase 2++ recommendations no longer matched the release year of these specifications, it was decided that the versions following Release 99 will be referenced according to a version number, Release 4 being the first new version reference.

The GSM recommendations are organized up to Release 99 in the following series

01 series: General;

02 series: Service Aspects;

03 series: Network Aspects;

04 series: MS-BS Interface and Protocols;

05 series: Physical Layer on the Radio Path;

06 series: Speech Coding Specification;

07 series: Terminal Adaptors for MSs;

08 series: BS-MSC Interface;

09 series: Network Interworking;

11 series: Equipment and Type Approval Specification;

12 series: Operation and Maintenance.

Each of the series contains a list of specifications identified by numbers. A given specification is therefore defined by its series number, followed by a recommendation number. For example, the 05.03 specification belongs to the physical layer on the radio path, and deals with channel coding issues.

In Release 4 of the 3GPP recommendations, the former GSM 01, 02, 03, 04, 05, and 06 series were kept for all GSM features that have not evolved with the third generation. From Release 4, these GSM series numbers were replaced by new series numbers, to be compliant with the 3GPP numbering. New series numbers can easily be deduced by adding 40 to the GSM series. Thus the 05 series describing the physical layer on the radio path become the 45 series from release 4. Also, the specification numbers in each series are deduced from the previous numbers by inserting a 0 as the first number. Thus, the R99 05.03 becomes the 45.003 from R4. Note that the different releases are continuously maintained.

The GPRS feature was introduced in Release 97 of the 3GPP recommendations. The GPRS recommendations are organized in three stages, as are all 3GPP recommendations:

- Stage 1: Description of GPRS services;
- Stage 2: Description of GPRS general architecture;
- Stage 3: Detailed description of different equipment implemented for GPRS with their external interfaces.

The Stage 1 GPRS recommendations describe the services that will be provided by GPRS. The service description is given in the 02 series recommendations for Release 97 and Release 98 of the 3GPP recommendations and in the 22 series recommendations from Release 99. The evolution of GPRS services is discussed in Working Group 1 (WG1) of TSG SA within the 3GPP.

The stage 2 GPRS recommendations describe the general architecture of GPRS, with nodes implemented in the network and the interface mechanisms between these nodes. The description of the architecture is given in the 03 series of the recommendations for Release 97 and Release 98 of the 3GPP recommendations and in 23 series recommendations from Release 99. The evolution of GPRS architecture is discussed in WG2 of TSG SA within the 3GPP.

The stage 3 GPRS recommendations describe in a detailed manner the equipment and its external interfaces implemented in the network for GPRS (see 04, 05, 06, and 08 series recommendations up to Release 99, and then 44, 45, and 46 series recommendations from Release 4). The evolution of the behavior of this equipment is discussed in working groups of several associated TSGs.

References

[1]   3GPP TS 22.105 Services and Service Capabilities (R99).

[2]   3GPP TS 22.060 General Packet Radio Service (GPRS); Stage 1 (R99).

Selected Bibliography

ITU-T recommendation F.700, "Framework Recommendation for Audio-Visual/Multimedia Services."

3GPP TS 23.060 Service Description; Stage 2 (R99).

3GPP TS 23.107 QoS Concept and Architecture (R99).

3

Overview of GPRS

GPRS represents an evolution of the GSM standard, allowing data transmission in packet mode and providing higher throughputs as compared with the circuit-switched mode. This evolution is usually presented under the designation of 2.5G to point out that it is a transition technology between 2G and 3G.

The GPRS network architecture reuses the GSM network nodes such as MSC/VLR, HLR, and BSS. New network nodes have been introduced for the transport of packet data. These nodes are the *gateway GPRS support nodes* (GGSN) and *serving GPRS support nodes* (SGSN). The subnetwork formed by the SGSNs and the GGSNs is called the GPRS core network. In order to reuse the GSM nodes, new interfaces have been defined between the GSM network nodes and the different elements of the GPRS core network. The GPRS logical architecture is described in Section 3.1.

The protocol layer has been split into two planes. On one side there is the transmission plane, which is mainly used for the transfer of user data. The signaling plane is used for the control and support of the transmission plane functions. Section 3.2 deals with the transmission and signaling planes.

GPRS has kept such main principles of the GSM radio interface as the notions of time slot, frame, multiframe, and hyperframe structures. It was indeed chosen by the operators and manufacturers involved in the system design to provide high-data-rate packet-switched services with minimized impacts on the GSM standard. The principles of the physical layer are given in Section 3.3.1. The details related to the physical layer are presented in Chapter 4.

One of the main GPRS characteristics is that a physical connection is established in uplink only when the MS needs to send continuous data to the network, and in downlink only when the network needs to send continuous data to the MS. This physical connection is released in one direction as soon as the sending entity has no more data to send. Different allocation schemes for *radio resource* (RR) management have been defined in order to multiplex several MSs on the same physical channel. An overview of the principles related to RR management is presented in Section 3.3.2. A complete description of RR management can be found in Chapter 5.

A logical entity called PCU has been introduced within the BSS to manage the GPRS functions over the radio interface. Section 3.4 deals with the BSS architecture and discusses the several possible locations of the PCU.

In a GPRS network, an RA identifies one or several cells. As soon as an MS enters a new RA, the network must be notified of this change in order to update its location. The SGSN is in charge of *GPRS mobility management* (GMM). An overview of GPRS mobility is proposed in Section 3.5. The details related to GMM are given in Chapter 7.

If data transmission in packet mode does not require the establishment of an end-to-end connection, it is necessary to establish a context between the mobile and the network in order to exchange packets. This context allows the network to identify the IP address of the MS, identify the access point with the external network, and define the QoS associated with data transmission in packet mode. The concept of PDP context is explained in Section 3.6. The details related to PDP context management are given in Chapter 7.

The BSS and the GPRS backbone network are connected via the Gb interface in order to exchange user data and signaling information. The principles of the Gb interface are given in Section 3.8. The details related to the Gb interface are given in Chapter 6.

When a context is established between the MS and the network, IP packet exchange may start at any time between the mobile and the network without establishing a connection beforehand. The packets are conveyed in the GPRS backbone network. An overview of the general architecture of the GPRS backbone network is presented in Section 3.9. A complete description of the user plane between the MS and external data packet network is given in Chapter 8.

3.1   GPRS Logical Architecture

Figure 3.1 shows the elements that are part of a GPRS network and their associated interfaces. A GPRS network is composed of the following network nodes:

- SGSN. The SGSN is the node that is serving the MS; it is responsible for GMM. It delivers packets to the MSs and communicates with the HLR to obtain the GPRS subscriber profile. It manages the registration of the new mobile subscribers in order to keep a record of their LA for routing purposes. The SGSN can be connected to one or several BSSs.

- GGSN. The GGSN provides interworking with external *packet data networks* (PDNs). It may be linked to one or several data networks. It is connected with SGSNs via an IP-based GPRS backbone network. The GGSN is a router that forwards incoming packets from the external PDN to the SGSN of the addressed MS. It also for-

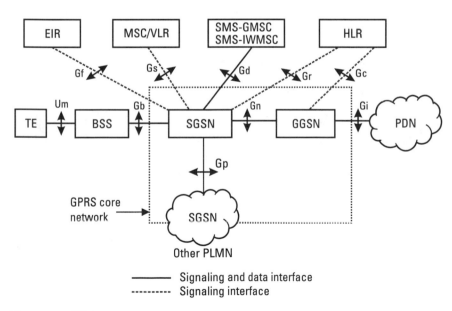

Figure 3.1   GPRS network architecture.

wards outgoing packets to the external PDN. The PDN is the external fixed data network to which is connected the GPRS network. An example of a PDN is the Internet network.

- HLR. The HLR is a database that contains, among other things, packet domain subscription data and routing information.

- *Mobile switching center/visitor location register* (MSC/VLR). The MSC coordinates the setting up of calls to and from GSM users and manages GSM mobility. The MSC is not directly involved in the GPRS network. It forwards circuit-switched paging for the GPRS-attached MSs to the SGSN when the Gs interface is present.

- BSS. The BSS ensures the radio connection between the mobile and the network. It is responsible for radio access management. The BSS is composed of two elements: the BTS and the BSC. The BTS integrates all the radio transmission and radio reception boards. The BSC is responsible for the management of the radio channels. The BSC has switching capabilities that are used for circuit-switched calls and can also be used for GPRS traffic.

- EIR. The EIR is a database that contains terminal identities.

3.2 Transmission and Signaling Planes

3.2.1 Transmission Plane

The transmission plane consists of a layered protocol structure providing user data transfer, along with associated procedures that control the information transfer such as flow control, error detection, and error correction. Figure 3.2 illustrates the layered protocol structure between the MS and the GGSN.

3.2.1.1 Air Interface

The air interface is located between the MS and the BSS. The protocols used on the air interface are as follows:

- *Radio link control/medium access control* (RLC/MAC). RLC provides a reliable radio link between the mobile and the BSS. MAC controls the access signaling procedures to the GPRS radio channel, and the multiplexing of signaling and RLC blocks from different users onto the GSM physical channel.

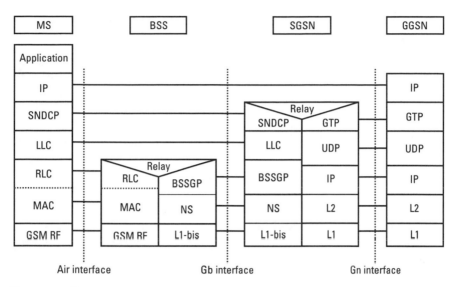

Figure 3.2 Transmission plane MS GGSN.

- GSM-RF layer. It is the radio subsystem that supports a certain number of logical channels. This layer is split into two sublayers: the *radio frequency layer* (RFL), which handles the radio and baseband part (physical channel management, modulation, demodulation, and transmission and reception of radio blocks), and the *physical link layer* (PLL), which manages control of the RFL (power control, synchronization, measurements, and channel coding/decoding).

A relay function is implemented in the BSS to relay the LLC PDUs between the air interface and the Gb interface.

3.2.1.2 Gb Interface

The Gb interface is located between the SGSN and the BSS. It supports data transfer in the transmission plane. The Gb interface supports the following protocols:

- *BSS GPRS protocol* (BSSGP). This layer conveys routing and QoS-related information between the BSS and SGSN.
- *Network service* (NS). It transports BSSGP PDUs and is based on a frame relay connection between the BSS and SGSN.

A relay function is implemented in the SGSN to relay the *packet data protocol* (PDP) PDUs between the Gb and Gn interfaces (IP PDUs in Figure 3.2).

3.2.1.3   Gn/Gp Interface

The Gn interface is located between two GSNs (SGSN or GGSN) within the same PLMN, while the Gp interface is between two GSNs in different PLMNs. The Gn/Gp interface is used for the transfer of packets between the SGSN and the GGSN in the transmission plane.

The Gn/Gp interface supports the following protocols:

- *GPRS tunnelling protocol* (GTP). This protocol tunnels user data between the SGSN and GGSN in the GPRS backbone network. GTP operates on top of UDP over IP. The layers L1 and L2 of the Gn interfaces are not specified in the GSM/GPRS standard.

- *User datagram protocol* (UDP). It carries GTP packet data units (PDUs) in the GPRS Core Network for protocols that do not need a reliable data link (e.g., IP).

- *Internet protocol* (IP). This is the protocol used for routing user data and control signaling within the GPRS backbone network.

3.2.1.4   Interface Between MS and SGSN

This interface supports the following protocols:

- *Subnetwork-dependent convergence protocol* (SNDCP). This protocol maps the IP protocol to the underlying network. SNDCP also provides other functions such as compression, segmentation, and multiplexing of network layer messages.

- *Logical link control* (LLC). This layer provides a highly reliable logical link that is independent of the underlying radio interface protocols. LLC is also responsible for the GPRS ciphering.

3.2.2   Signaling Plane

The signaling plane consists of protocols for control and support of the transmission plane functions. It controls both the access connections to the GPRS network (e.g., GPRS attach and GPRS detach) and the attributes of an established network access connection (e.g., activation of a PDP address), manages the routing of information for a dedicated network connection in order to support user mobility, adapts network resources depending on the QoS parameters, and provides supplementary services.

3.2.2.1 Between MS and SGSN

Figure 3.3 shows the signaling plane between the MS and the SGSN. This plane is made up of the following protocols:

- GMM. The GMM protocol supports mobility management functionalities such as GPRS attach, GPRS detach, security, RA update, and location update (see Section 3.5).

- *Session management* (SM). The SM protocol supports functionalities such as PDP context activation, PDP context modification, and PDP context deactivation (see Section 3.6).

3.2.2.2 Between Two GSNs

In the signaling plane, Gn/Gp interfaces are used for the transfer of signaling between the GSNs in the GPRS backbone network. Figure 3.4 shows the signaling plane between two GSNs.

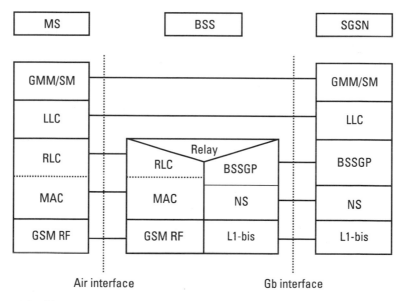

Figure 3.3   Signaling plane MS-SGSN.

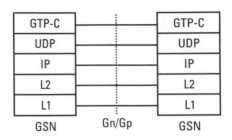

Figure 3.4   Signaling plane GSN-GSN.

The signaling plane between two GSNs is made up of the following protocols:

- GTP for the control plane (GTP-C). This protocol tunnels signaling messages between SGSNs and GGSNs, and between SGSNs, in the GPRS core network (see Section 3.9).
- UDP. This protocol transfers signaling messages between GSNs.

3.2.2.3   Interface with Signaling System No. 7

The various GSNs of the GPRS backbone network use a *Signaling System No. 7* (SS7) network to exchange information with GSM SS7 network nodes such as HLR, MSC/VLR, EIR, and SMS-GMSC. The SS7 network provides facilities to quickly exchange messages between GPRS backbone network nodes irrespective of data transmission through the GPRS PLMN network.

In the GSM/GPRS backbone network, we found the following protocols for SS7 signaling:

- *Message transfer part* (MTP). The three MTP layers allow signaling messages to be exchanged through the SS7 network.
- *Signaling connection control part* (SCCP). The SCCP protocol layer allows the service to be used in connected mode and messages to be exchanged between different PLMNs by using an international gateway for SS7 address translation between an SS7 global address (based on the E.164 numbering plan) and an SS7 local address.
- *Transaction capabilities application part* (TCAP). The TCAP protocol layer allows dialogs to be structured in an independent manner from any application.

Table 3.1
New Interfaces with the SS7 Network

Interface Name	Localization	Mandatory/Optional
Gr	SGSN-HLR	Mandatory
Gc	GGSN-HLR	Optional
Gf	SGSN-EIR	Optional
Gd	SGSN-SMS GMSC or SGSN-SMS IWMSC	Optional
Gs	SGSN-MSC/VLR	Optional

- *Mobile application part* (MAP). The MAP protocol layer allows mobile mobility to be managed within different equipment nodes of the NSS across SS7 networks.

As new equipment nodes have been introduced in GSM networks to support the GPRS feature, new interfaces were defined with the HLR, MSC/VLR, EIR, and SMS-GMSC. Table 3.1 lists the new interfaces with SS7 network.

Gr Interface

The Gr interface is defined between the SGSN and HLR. It allows the SGSN to retrieve or update GPRS subscription and GPRS location information in the HLR during location-management or authentication procedures. The MAP protocol has been modified to take into account this interface. Figure 3.5 shows the signaling plane on the Gr interface.

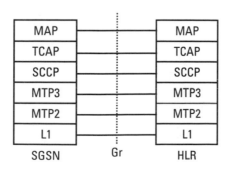

Figure 3.5   Signaling plane on the Gr interface.

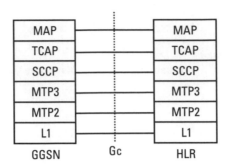

Figure 3.6   Signaling plane on the Gc interface.

Gc Interface

The Gc interface is defined between the GGSN and the HLR. The GGSN contacts the HLR in order to determine the SGSN address where the MS is located and if the MS is reachable. The MAP protocol has been modified to take into account this interface. Figure 3.6 shows the signaling plane on the Gc interface. (*Note:* If a GGSN does not have a SS7 MAP interface, it will interface to a GSN performing a GTP-MAP protocol-conversion in order to retrieve the needed information from the HLR via the Gc interface.)

Gf Interface

The Gf interface is defined between the SGSN and EIR. It is used by the SGSN to contact the EIR database during the identity check procedure. It allows the SGSN to check the IMEI against the EIR. The MAP protocol has been modified to take into account this interface. Figure 3.7 shows the signaling plane on the Gf interface.

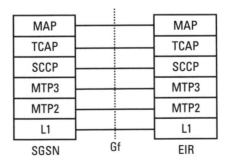

Figure 3.7   Signaling plane on the Gf interface.

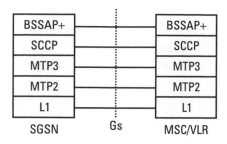

Figure 3.8   Signaling plane on the Gs interface.

Gs Interface

The Gs interface is defined between the MSC/VLR and the SGSN. It allows for the coordination of circuit-switched and packet-switched paging in the SGSN as well as location information of any MS attached to both circuit and packet services. This interface is only present in a network that operates in network operation mode I (see definition in Section 3.5.3.1). The BSS application part+ (BSSAP+) allows mobility functionality to be managed on the Gs interface. Figure 3.8 shows the signaling plane on the Gs interface.

Gd Interface

The Gd interface is defined between an SGSN and an SMS-GMSC or an SMS-IWMSC. The progress of a short message intended for delivery to an MS requires in circuit and packet modes a gateway function *Short Message Service-gateway MSC* (SMS-GMSC) between the mobile network and the network that provides access to the SMS center. An SMS to be delivered to an MS is routed from the SMS-GMSC toward the SGSN on the Gd interface if this SMS is to be sent over GPRS.

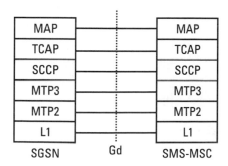

Figure 3.9   Signaling plane on the Gd interface.

The progress of a short message originated by the MSs requires in circuit and packet modes a PLMN interworking server SMS-IWMSC (interworking MSC) that provides access to the SMS center. An SMS originated by an MS is routed from the SGSN toward the SMS-IWMSC on the Gd interface if the SMS is to be sent over GPRS.

The MAP protocol has been updated as a consequence of the signaling exchange between the SGSN and the SMS-GMSC or the SMS-IWMSC. Figure 3.9 shows the signaling plane on the Gd interface.

3.3   Radio Interface

3.3.1   Physical Layer Principles

3.3.1.1   Packet Data Channel

The GPRS physical layer is based on that of the GSM (see Chapter 1). The access scheme is TDMA, with eight basic physical channels per carrier (TS 0 to 7).

A physical channel uses a combination of frequency- and time-division multiplexing and is defined as a radio frequency channel and time slot pair. The physical channel that is used for packet logical channels is called a *packet data channel* (PDCH). PDCHs are dynamically allocated in the cell by the network. The PDCH is mapped on a 52-multiframe, as shown in Figure 3.10. The 52-multiframe consists of 12 radio blocks (B0 to B11) of 4 consecutive TDMA frames and 4 idle frames (frames 12, 25, 38, and 51), amounting to a total of 52 frames.

3.3.1.2   Packet Data Logical Channel

GPRS, like GSM, uses the concept of logical channels mapped on top of the physical channels. Two types of logical channels have been introduced, namely traffic channels and control channels. Three subtypes of control channels have been defined for GPRS: broadcast, common control, and associated. In addition, the GSM common control channels (BCCH, CCCH, and RACH) may be used to access the network and establish packet transfer.

The different packet data logical channels are:

- *Packet broadcast control channel* (PBCCH). The presence of PBCCH in the cell is optional. The PBCCH broadcasts information relative to the cell in which the mobile camps and information on the

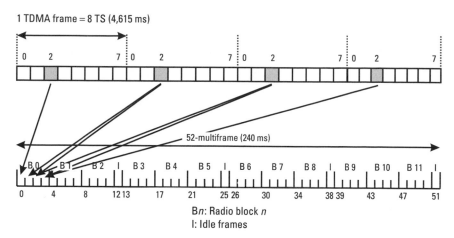

1 TDMA frame = 8 TS (4,615 ms)

52-multiframe (240 ms)

Bn: Radio block *n*
I: Idle frames

Figure 3.10 Time slots and TDMA frames.

neighbor cells. This information is used by the mobile in order to access the network. When there is no PBCCH in the cell, the information needed by the mobile to access the network for a packet transfer is broadcast on BCCH.

- *Packet common control channel* (PCCCH). The PCCCH is present in the cell only if the PBCCH is present in the cell. When it is not present in the cell, the common control signaling for GPRS is handled on the GSM *common control channels* (CCCH). PCCCH is composed of *packet random access channel* (PRACH), used for random access, *packet paging channel* (PPCH), used for paging, and *packet access grant channel* (PAGCH), used for access grant. The PRACH is used by the MS to initiate uplink access to the network. The PPCH is used by the network to page the mobile in order to establish a downlink packet transfer. The PAGCH is used by the network to assign radio resources to the mobile for a packet transfer.

- *Packet data traffic channel* (PDTCH). The PDTCH is used to transfer user data during uplink or downlink packet transfer. The PDTCH is a unidirectional channel, either uplink (PDTCH/U) for a mobile-originated packet transfer or downlink (PDTCH/D) for a mobile-terminated packet transfer. A PDTCH is a resource allocated on one physical channel by the network for user data transmission.

Table 3.2
GPRS Logical Channels

Logical Channel	Abbreviation	Uplink/ Downlink	Task
Packet broadcast control channel	PBCCH	DL	Packet system broadcast information
Packet paging channel	PPCH	DL	MS paging for downlink transfer establishment
Packet random access channel	PRACH	UL	MS random access for uplink transferestablishment
Packet access grant channel	PAGCH	DL	Radio resources assignment
Packet timing advance control channel	PTCCH	UL/DL	Timing advance update
Packet associated control channel	PACCH	UL/DL	Signaling associated with data transfer
Packet data traffic channel	PDTCH	UL/DL	Data channel

- *Packet associated control channel* (PACCH). The PACCH is a unidirectional channel that is used to carry signaling during uplink or downlink packet data transfer. The uplink PACCH carries signaling from the MS to the network and the downlink PACCH carries signaling from the network to the mobile. The PACCH is dynamically allocated on a block basis.

- *Packet timing advance control channel* (PTCCH). The PTCCH is a bidirectional channel that is used for TA update. The PTCCH is an optional channel. The PTCCH when present is mapped on frames number 12 and 38 of the 52–multiframe.

Table 3.2 lists the various GPRS logical channels.

3.3.1.3　Multislot Classes Definition

In order to provide higher throughputs, a GPRS MS may transmit or receive in several time slots of the TDMA frame. The multislot capability is indicated by the multislot class of the GPRS MS. This multislot class is defined by several parameters such as the maximum number of time slots supported by the MS per TDMA frame in uplink and in downlink. The multislot class of the MS is sent to the network during the GPRS attach procedure. A detail of multislot classes is given in Section 4.2.1.

3.3.1.4 Cell Reselection

In GPRS as in GSM, the mobile performs cell reselection. However, there are some differences compared with GSM.

In GPRS the mobile performs cell reselection when it is in idle mode but also during packet transfer. The cell reselection is either performed by the mobile autonomously or optionally controlled by the network. Unlike GSM, there is no handover in GPRS but only cell reselections. So when there is a reselection during a packet transfer, this latter is interrupted and it has to be started again in the new cell. There is an interruption of the packet transfer during the reselection phase.

Although the GPRS cell reselection algorithms used by the mobile are based on the same principles as those used in GSM, they have been slightly modified in order to provide more flexibility. These algorithms are described in Section 5.3.

3.3.1.5 Radio Environment Monitoring

The MS performs different types of radio measurements that are reported to the network and used by it for RLC. These estimations are also used by the mobile itself to compute its transmission power (open-loop power control; see Section 4.1.3.1), for cell selection and cell reselection.

The mobile performs the following types of measurement:

- *Received signal level (RXLEV) measurements.* The RXLEV measurements are performed on both serving cell and neighbor cells for the purpose of cell reselection. During packet-transfer mode (see Section 3.3.2.1), the serving cell RXLEV measurement can also be used for downlink coding scheme adaptation (see Section 4.1.2.1), network-controlled cell reselection, and downlink and uplink power control.

- *Quality (RXQUAL) measurements.* The RXQUAL is computed from the average BER before channel decoding. During packet-transfer mode the mobile estimates the quality of the downlink blocks it receives. In packet idle mode, no quality measurements are performed. The RXQUAL can be used by the network for network-controlled cell reselection, dynamic coding scheme adaptation, and downlink power control. The RXQUAL is the current GSM quality indicator.

- *Interference measurements.* These measurements have been introduced for GPRS. They correspond to a received signal level measurement performed on a frequency that is different from a beacon frequency. The interest is in having an estimation of the interference level on the PDTCH. It can be used by the network to optimize the mobile RR allocation, to select a more appropriate coding scheme, to trigger a network-controlled cell reselection, and for power control or for network statistics.

The measurements are either used by the MS for its own purposes or by the network. The RXLEV and RXQUAL measurements are also performed at the BSS side for each MS. They are used for network-controlled cell reselection, uplink power control (closed loop, see Section 4.1.3.1), and dynamic coding scheme adaptation.

3.3.1.6 Principles of Power Control

Power control can be used in order to improve the spectrum efficiency while maintaining radio link quality and reducing the power consumption in the MS. Power control in GPRS is more complicated than for a circuit-switched connection, since there is not necessarily a continuous two-way connection.

Note that power control can be performed in both uplink and downlink directions. Uplink power control and downlink power control are described in Sections 4.1.3.1 and 4.1.3.2, respectively.

3.3.1.7 Channel Coding

The GPRS user data is sent on radio blocks encoded with one of four channel coding schemes (CS1, CS2, CS3, CS4). The GPRS signaling is sent with the CS1 channel coding scheme. This scheme provides the highest protection level against error transmission, while the CS4 channel coding scheme

Table 3.3
Throughput Associated with Coding Scheme

Coding Scheme	Throughput (Kbps)
CS-1	9.05
CS-2	13.4
CS-3	15.6
CS-4	21.4

provides the lowest protection level. The more the channel coding scheme provides an efficient protection level against error transmission, the more the useful data throughput decreases due to redundant information added to source data. The channel coding scheme used on uplink and downlink depends on the radio quality between the network and the MS. Table 3.3 gives the throughputs associated with each coding scheme.

3.3.2   RR Management Principles

3.3.2.1   RR Operating Modes

At the RR level, the MS behavior is dependent on two operating RR states. These states, packet idle mode and packet transfer mode, allow the RR activity of the MS to be characterized.

Packet Idle Mode

When the MS is in packet idle mode, no radio resources are allocated. Leaving packet idle mode occurs when upper layers request the transfer of uplink data requiring the assignment of uplink resources from the network. It also occurs at the reception of a downlink resource assignment command from the network for a downlink transfer.

In case of downlink transfer, the mobile switches from packet idle mode to packet transfer mode when it receives the downlink assignment command from the network. In the case of uplink transfer, the mobile leaves packet idle mode when it requests the assignment of uplink resources to the network. However, switching to packet transfer mode is not instantaneous. The mobile switches to packet transfer mode only when it has been uniquely identified at the network side; this will be explained in more detail in Chapter 5. Thus there is a period between packet idle mode and packet transfer mode during which the mobile is in a transitory state.

During packet idle mode, the MS listens to its PCH and the CBCH. This last one is the PBCCH when present in the cell; otherwise it is the BCCH.

Packet Transfer Mode

When the MS is in packet transfer mode, it is clearly identified at the network side and uplink or/and downlink radio resources are allocated.

Switching from packet transfer mode to packet idle mode occurs when the network releases all downlink and uplink resources. This transition can also occur in the case of an abnormal condition during packet transfer mode

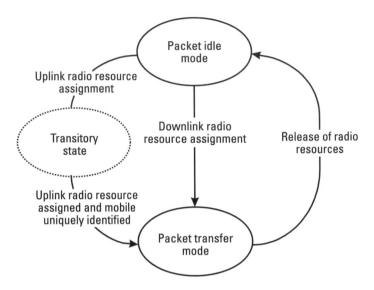

Figure 3.11 Transition between RR operating modes.

(e.g., radio link failure) or when the mobile decides on a cell reselection toward a new cell.

During packet transfer mode, the mobile transmits and receives data. Figure 3.11 summarizes the transition between the different RR states.

3.3.2.2 Temporary Block Flow

A *temporary block flow* (TBF) is a physical connection between the RR entity in the MS and the RR entity at the network side to support the unidirectional transfer of LLC protocol data units over PDCH. A TBF is characterized by one or several PDCHs allocated by the network to an MS for the duration of the data transfer. Once the data transfer is finished, the TBF is released.

When the mobile must send continuous data to the network, it requests the establishment of an uplink TBF by sending signaling information over CCCH or PCCCH. When the network wants to send data to the mobile, it assigns a downlink TBF between the two RR entities.

A downlink TBF supports the transfer of data from the network to the mobile, while an uplink TBF supports the transfer of data from the mobile to the network. One uplink and one downlink TBF can be supported at the same time between the two RR entities. These two TBFs are defined as concurrent TBFs.

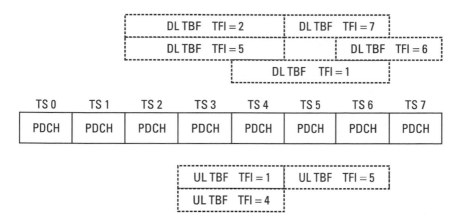

Figure 3.12 Example of mapping of TBFs with their respective TFI onto PDCHs.

The number of TBFs per mobile and per direction is limited to one. However TBFs belonging to different mobiles can share the same PDCH.

Note As a packet transfer session is composed of a lot of requests, responses, and acknowledgments, many consecutive uplink and downlink TBFs are established for the same session (e.g., Web browsing).

Each TBF is identified by a *temporary flow identifier* (TFI) assigned by the network. So in case of concurrent TBFs, one TFI identifies the uplink TBF and another one the downlink TBF. The TFI is used to differentiate TBFs sharing the same PDCHs in one direction.

Figure 3.12 gives an example of TBF mapping for different mobiles onto PDCHs. A downlink TBF, identified by a TFI equal to 2, has assigned resources on the PDCH numbers 2, 3, and 4, while another downlink TBF identified by a TFI equal to 1 has assigned resources on the PDCH numbers 4, 5, and 6. An uplink TBF identified by a TFI equal to 4 has assigned resources on the PDCH numbers 3 and 4.

3.3.2.3 Allocation Modes on the Uplink

Several MSs may be multiplexed on the same PDCH. In order to share the uplink bandwidth between several mobiles mapped on the same PDCH, different allocation schemes have been defined to allocate an uplink radio block instance to a particular mobile.

Three allocation schemes exist for medium access control:

- Dynamic allocation;
- Extended dynamic allocation;
- Fixed allocation.

The second scheme is optional for the mobile while the others are mandatory. On the network side, either fixed allocation or dynamic allocation must be implemented. Extended dynamic allocation is optional for the network.

Dynamic Allocation

In principle, dynamic allocation allows uplink transmission to mobiles sharing the same PDCH, on a block-by-block basis. During the uplink TBF establishment, an *uplink state flag* (USF) is given to the MS for each allocated uplink PDCH. The USF is used as a token given by the network to allow transmission of one uplink block.

Whenever the network wants to allocate one radio block occurrence on one uplink PDCH, it includes, on the associated downlink PDCH, the USF in the radio block immediately preceding the allocated block occurrence. When the mobile decodes its assigned USF value in a radio block sent on a downlink PDCH associated with an allocated uplink one, it transmits an uplink radio block in the next uplink radio block occurrence, that is, the $B(x)$ radio block if the USF was detected in the $B(x-1)$ radio block.

The principle of dynamic allocation is illustrated in Figure 3.13.

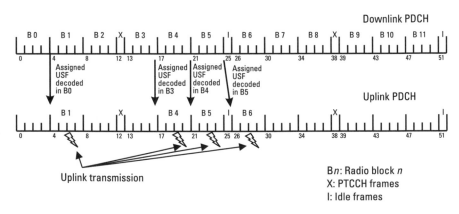

Figure 3.13 Principle of dynamic allocation.

The USF is included in the header of each downlink RLC/MAC block. The USF coding (3 bits) enables eight mobiles to be multiplexed on the same uplink PDCH.

Dynamic allocation implies the constant monitoring (radio block decoding) of the downlink PDCHs associated with the allocated uplink PDCHs.

As explained previously, the USF allows the sending of one block in the next uplink occurrence. However, dynamic allocation can also be used in such a way that the decoding of one USF value allows the mobile to send four consecutive uplink blocks on the same PDCH. The choice between one block or four blocks is indicated during the TBF establishment by the network to the mobile.

The concept of USF granularity has been introduced in order to indicate the number of uplink radio blocks to be sent upon detection of the assigned USF by the MS. The USF granularity is conveyed to the MS during the uplink TBF establishment. Thus it involves the MS transmitting either a single radio block or a sequence of four consecutive radio blocks starting on the B(x) radio block if the USF was detected in the B($x - 1$) radio block.

Extended Dynamic Allocation

The extended dynamic allocation scheme offers an improvement over the dynamic allocation scheme. Some RR configurations are not compliant with all MS multislot classes in the dynamic allocation scheme. In the dynamic allocation scheme, the MS must decode all USF values on all downlink PDCHs associated with the allocated uplink PDCHs.

The mobile monitors its assigned PDCHs starting from the lowest numbered one (the one that is mapped on the first allocated time slot in the TDMA frame), then it monitors the next lowest numbered time slot, and so on. Whenever the MS detects its assigned USF value on a PDCH, it transmits one radio block or a sequence of four radio blocks on the same PDCH and all higher-numbered assigned PDCHs. The mobile does not need to monitor the USF on these higher PDCHs. This is of particular interest in some RR configurations that are not compliant with all MS multislot classes in the dynamic allocation scheme.

Let us take the example of a class 12 MS, which is defined by:

- A maximum number of four receive time slots per TDMA frame;
- A maximum number of four transmit time slots per TDMA frame;
- A total number of transmit and receive time slots per TDMA frame less than or equal to five.

The network cannot allocate four uplink PDCHs to a MS multislot class 12 with the dynamic allocation. Indeed, the MS must decode the USF fields on the four associated downlink PDCHs. This means that the MS would have to receive on four time slots to be able to transmit on four time slots. That gives a total number of eight received and transmit time slots, which is not compliant with a multislot class 12 MS. In the case of extended dynamic allocation, the network can allocate four uplink PDCHs without exceeding a total number of five receive and transmit time slots.

Fixed Allocation

Fixed allocation enables a given MS to be signaled predetermined uplink block occurrences on which it is allowed to transmit. The network assigns to each mobile a fixed uplink resource allocation of radio blocks onto one or several PDCHs.

The network allocates uplink radio blocks using bitmaps (series of zeros and ones). A 0 indicates that the mobile is not allowed to transmit, and a 1 indicates a transmission occurrence. The bitmaps are sent during the establishment of the uplink TBF. If more uplink resources are required during the uplink TBF, the network sends a bitmap in downlink on the PACCH.

A fixed allocation TBF operates as an open-ended TBF when an arbitrary number of octets are transferred during the uplink TBF. When the allocated bitmap ends, the MS requests a new bitmap if it wishes to continue the TBF.

A fixed allocation TBF operates as a close-ended TBF when the MS specifies the number of octets to be transferred during the uplink TBF establishment.

Comparison of Allocation Schemes

Fixed allocation allows for an efficient usage of the MS multislot capability, as the downlink monitoring is limited to one time slot (listening of the PACCH). The advantage of dynamic allocation is that the management of uplink resources is much more flexible.

From an implementation point of view, dynamic allocation is easier to implement on the network side. An efficient management of bitmap in fixed allocation is not so easy compared with USF handling. In fact, the management of resource allocation with anticipation (as needed in fixed allocation) is made more complex by the bursty nature of packet transfer. Most of the time the duration of the uplink TBF is completely unknown at the network side at the beginning of the TBF. This is why it is very difficult to manage radio block allocation at the beginning of the TBF. If the bitmap allocation is too short, the reaction time of the network for reallocation of new bitmaps will increase the duration of the TBF. If the bitmap allocation is too large compared with the TBF duration, block bitmap will be reallocated at the network side in order to avoid waste of uplink resources. Dynamic allocation is very easy to manage, since the USF allocation needs to anticipate only a few block periods.

3.4 BSS Architecture

3.4.1 PCU

In order to introduce GPRS within the BSS, the PCU concept has been defined. The PCU stands for a logical entity that manages packet toward the radio interface. The PCU communicates with the *channel codec unit* (CCU), positioned in the BTS.

The PCU is in charge of RLC/MAC functions such as segmentation and reassembly of LLC frames, transfer of RLC blocks in acknowledged or unacknowledged mode, radio resource assignment, and radio channel management. The CCU handles GSM layer 1 functions such as channel decoding, channel encoding, equalization, and radio channel measurements.

As shown in Figure 3.14, the PCU can be located either at the BTS site, the BSC site, or the SGSN site. When the PCU is located at the BSC or SGSN site, it is referred to as being a remote PCU.

If the PCU is located at the BSC, it could be implemented as an adjunct unit to the BSC. When the PCU is located at the SGSN side, the BSC is transparent for frames transmitted between the PCU and the CCU. This PCU location implies the implementation of a signaling protocol between the BSC and the PCU (e.g., for time slot management, access on CCCH management). A protocol is also needed when the PCU is located at the BTS side.

In the case of remote PCU, GPRS traffic between the PCU and the CCU are transferred through the Abis interface. The Abis interface in GSM is based on TRAU frames carrying speech data and having a fixed length of 320 bits (every 20 ms). This corresponds to a throughput of 16 Kbps per Abis channel.

As the PCU is supporting functions such as RLC block handling (retransmission, segmentation, and so forth) and access control, it needs to

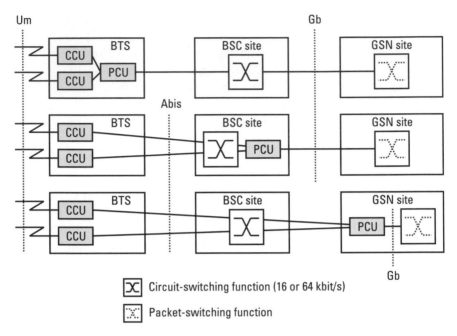

Figure 3.14 Remote PCU position. (*From:* [1].)

know the GSM radio interface timing. This implies in the case of a remote PCU the design of a synchronous interface between the PCU and the CCU. The PCU must be able to determine in which radio frames is sent an RLC/ MAC block. In case of a PCU located at the BTS side, the PCU easily knows the radio interface time.

The remote PCU solution requires the sending of in-band information between the PCU and the CCU (for transmission power indication, channel coding indication, and synchronization between the CCU and the PCU).

The advantages and drawbacks of each solution are listed next.

PCU at the BTS Side

Advantages:

- There is an internal interface between the PCU and the CCU.
- There is a low round-trip delay (the round-trip delay is the time between the transmission of a block and the reception of the answer).

- There is no waste of Abis bandwidth due to retransmission of RLC blocks.

Drawbacks:

- There is a likely impact on the existing BTS hardware (when migrating from a circuit-switched network to a GPRS one). This is an important drawback considering the number of BTSs in the field.

PCU at the BSC Side

Advantages:

- There is an internal interface between the PCU and the BSC.

Drawbacks:

- There are likely hardware impacts on the current BSC (however, the number of BSCs is lower compared with the number of BTSs).
- There is a greater round-trip delay.
- A synchronous protocol is needed between the PCU and the CCU.
- Abis bandwidth is wasted in case of RLC blocks retransmission.

PCU at the SGSN Side

Advantages:

- There is no hardware impact on the current GSM network (BTS, BSC).
- There is a smooth introduction of GPRS in the network.

Drawbacks:

- There is a greater round-trip delay (longer TBF establishment).
- A synchronous protocol is needed between the PCU and the CCU.
- Abis bandwidth is wasted in case of RLC blocks retransmission.
- A protocol is needed between the BSC and the PCU.

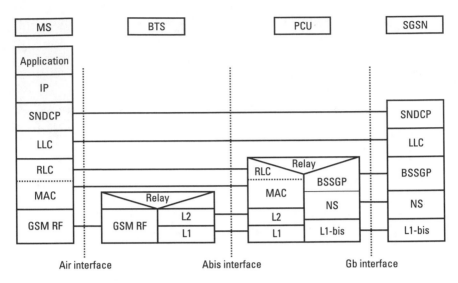

Figure 3.15 BSS transmission plane for a remote PCU.

3.4.2 Transmission Plane

When the PCU function is not implemented in the BTS, a new protocol (L1/L2) between the PCU and the BTS is introduced. This protocol ensures transmission of RLC/MAC blocks from the PCU to the CCU, and allows in-band signaling for control of the CCU from the PCU and synchronization between both entities. Figure 3.15 shows the BSS transmission plane for a remote PCU.

When the PCU is located in the BTS, a protocol (L1/L2/L3) is needed between the BSC and the BTS for the transmission of the LLC frames. The BSS transmission plane is shown in Figure 3.16.

3.4.3 Signaling Plane

Depending on the PCU location, the signaling plane within the BSS will not be the same.

3.4.3.1 PCU in the BTS

The RR layer is at the BTS side. The (L1/L2/L3) protocol is used for the transmission of signaling between the BSC and the BTS. Figure 3.17 shows the signaling plane when the PCU is located in the BTS.

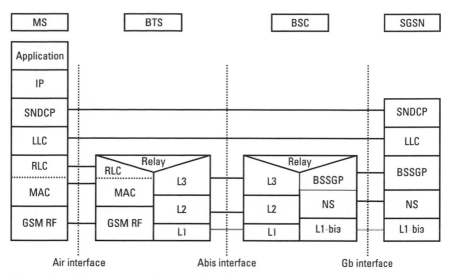

Figure 3.16 BSS transmission plane when the PCU is in the BTS.

3.4.3.2 PCU at BSC or SGSN Side

The signaling plane implies the usage of the (L1/L2) protocol for RR signaling transmission. Figure 3.18 shows the signaling plane when the PCU is located at BSC side or SGSN side.

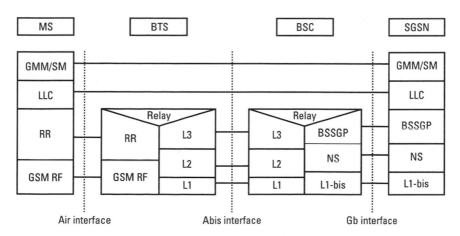

Figure 3.17 Signaling plane when PCU is in the BTS.

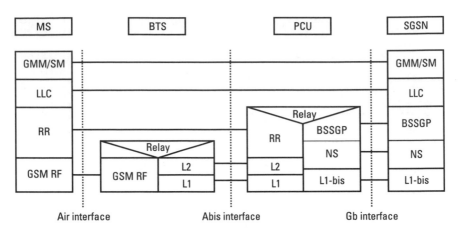

Figure 3.18 Signaling plane when PCU is at BSC or SGSN side.

3.5 Mobility

3.5.1 RA

A PLMN network supporting GPRS is divided into RAs. Each RA is defined by the operator of the PLMN network and may contain one or several cells. A LA is a group of one or several RAs. The RA defines a paging area for GPRS, while the LA defines a paging area for incoming circuit-switched calls. Actually, when the network receives an incoming call for a mobile not localized at cell level but localized at RA level, it broadcasts a paging on every cell belonging to this RA. The RA concept is illustrated in Figure 3.19.

If the MS moves to a new LA, it also moves to a new RA. Each RA is identified by a *routing area identifier* (RAI). This is made up of a *location area identifier* (LAI) and a *routing area code* (RAC). Figure 3.20 gives the structure of the RAI.

The LAI identifies the LA, with the *mobile country code* (MCC) indicating the PLMN country, the *mobile network code* (MNC) identifying the PLMN network in this country, and the *location area code* (LAC) identifying the LA.

The RAI of each RA is broadcast on all cells belonging to this RA. This way, the MS is able to detect a new RA by comparing the RAI it had previously saved with the one broadcast in the new cell, and then to signal to the network its RA change.

The MS may also signal to the network the RA in which it is located upon expiry of a periodic timer. This procedure allows the network to check

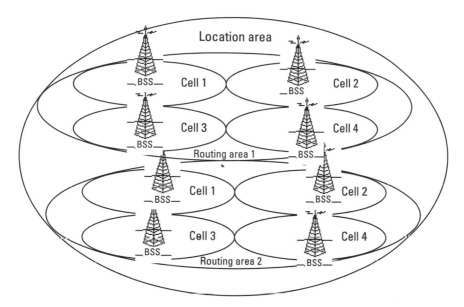

Figure 3.19 RA concept.

that the MS is within coverage of its RA. Owing to this procedure, the network knows if it may continue to route incoming calls for this MS toward this RA.

When an MS attached for circuit and packet services detects a new LA on the serving cell after having changed the cell, it will signal to the network its LA and RA change.

3.5.2 GMM States

Three global states are defined for GPRS mobility at the GMM layer level. These global states, GMM IDLE, STANDBY, and READY allow for characterization of the GMM activity of a GPRS mobile. They are managed in the MS and in the SGSN for each MS, and the transitions between states are

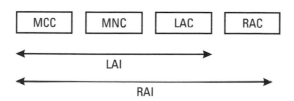

Figure 3.20 Structure of RAI.

slightly different on the MS and SGSN sides. A GPRS mobile is in GMM IDLE state when it is not attached for GPRS service. In this state, there is no GPRS mobility context established between the MS and the SGSN; this means that no information related to the MS is stored at SGSN level. In GMM STANDBY and READY states, a GPRS mobility context is established between the MS and the SGSN. A GPRS mobile is in GMM STANDBY state when it is attached for GPRS services and its location is known by the network at the RA level. A GPRS mobile is in GMM READY state when it is attached for GPRS services and its location is known by the network at the cell level.

A GPRS mobile goes to GMM IDLE state when it has just detached from GPRS. The SGSN goes to GMM IDLE state for a given MS upon

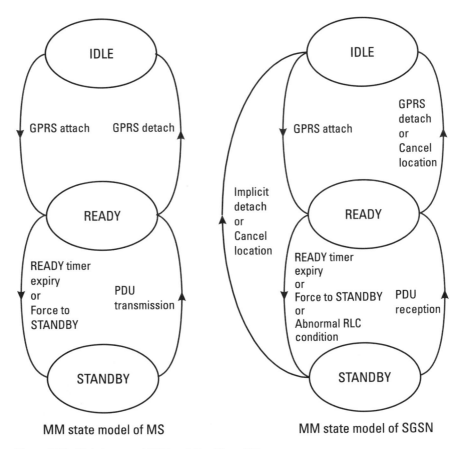

Figure 3.21 Global states of GPRS mobility. (*From:* [1].)

receipt of the GPRS detach message, upon implicit detach when no MS activity is detected, or upon receipt of cancel location from HLR for operator purposes.

A GPRS mobile goes to GMM READY state when it has just sent a packet to the network. For each packet sent to the network, the MS reinitializes a READY timer. The SGSN goes to GMM READY state for a given MS when it receives an LLC PDU from it. For each LLC PDU received from the MS, the SGSN reinitializes a READY timer related to the MS.

A GPRS mobile goes to GMM STANDBY state from GMM READY state either upon expiry of the READY timer, or upon the receipt of an explicit request from the SGSN to force the GMM STANDBY state. The SGSN goes to GMM STANDBY state for one given MS either upon expiry of the READY timer, or upon explicit request from the network to force the GMM STANDBY state, or on an irrecoverable disruption of a radio transmission found at RLC layer level.

Note   The network may force the GMM STANDBY state in order to reduce the signaling load in the network. In fact, as explained below, the MS has to perform several procedures in GMM READY state that are not authorized in GMM STANDBY state, such as notification of cell change. Furthermore, the GMM STANDBY state enables optimization of MS autonomy, since the MS need not send as much signaling over the air interface as compared with the GMM READY state.

Figure 3.21 shows the transitions between the three GMM states.

Table 3.4 summarizes the list of authorized procedures relative to GMM states.

Table 3.4
Authorized Procedures Relative to GMM States

	IDLE	STANDBY	READY
PLMN selection	Yes	Yes	Yes
GPRS cell (re)selection	No	Yes	Yes
GPRS packet transfer	No	Yes	Yes
Paging	No	Yes	No
Routing area update procedure	No	Yes	Yes
Notification of cell change	No	No	Yes
Radio link measurement reporting	No	No	Yes

The three global states lead to different behaviors of the MS at the radio interface level. They are therefore sent to the RR management layer of the MS.

3.5.3   Overview of GMM Procedures

3.5.3.1   Paging

The network may page an MS for circuit-switched and packet-switched services. These two services are managed in the backbone network by two different nodes: the MSC for routing of circuit-switched calls and the SGSN for routing of packet-switched calls. If there is no paging coordination between the circuit-switched and packet-switched services, the paging for circuit-switched and packet-switched services will not necessarily arrive at the MS on the same logical channel over the radio interface. This implies that the MS has to simultaneously monitor several logical channels for paging detection, a difficult task for MS receivers. In order to ease the MS behavior with respect to paging detection, paging coordination between circuit-switched and packet-switched services may be implemented in the network by adding a new interface, called the Gs interface, between the MSC and SGSN. This interface enables an incoming circuit-switched call to be routed from the MSC to the SGSN; this will allow the mobile to detect the circuit-switched and packet-switched services in the same logical channel.

Paging modes are defined by the recommendations to allow different paging implementations in the network. These paging modes take into account parameters such as the paging coordination method between circuit-switched and packet-switched services and the presence or absence of PCCCH paging channels. The paging mode is broadcast by the network on each GPRS cell.

Three *network modes of operation* (NMOs) are defined for paging:

Mode I. Circuit-switched paging messages are sent on the same PCHs as packet-switched paging, since paging coordination is supported (i.e., on PCCCH paging channels if allocated in the cell, or otherwise on CCCH paging channels).

Mode II. Packet-switched paging messages are sent on CCCH paging channels.

Mode III. Circuit-switched paging messages are sent on CCCH paging channels and packet-switched paging messages are sent on PCCCH

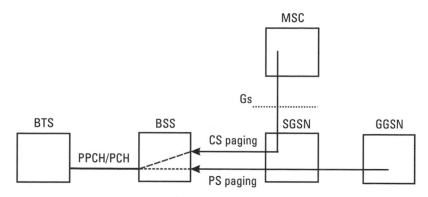

Figure 3.22 Paging in idle mode with CCCH/PCCCH for network operation mode I.

paging channels if they exist in the cell, or on CCCH paging channels otherwise. Mode III is thus equivalent to mode II if the PCCCH paging channels are not present in the cell.

Mode I

The network operates in mode I when the Gs interface is present between the MSC and the SGSN. In this mode, the network sends the paging messages on the same logical channels for circuit-switched and packet-switched services. When an MS both IMSI- and GPRS-attached is in idle mode, it monitors for any kind of incoming calls (circuit-switched and GPRS) the PCCCH channels if they are present, and the CCCH channels otherwise. Figure 3.22 illustrates a paging in idle mode when the network operates in mode I.

In mode I, when an MS both IMSI- and GPRS-attached is in packet transfer mode, the circuit-switched paging occurs on the PACCH of the TBF in progress. Figure 3.23 illustrates a paging in packet transfer mode when the network operates in mode I.

Mode II

In network operation mode II, the Gs interface is not present between the MSC and the SGSN, and the PPCHs are not present in the cell. In this mode the network sends paging messages for circuit-switched and packet-switched services on CCCH paging channels.

In the case of a packet transfer in progress, the MS may receive a circuit-switched paging on the CCCH paging channels. The MS then has the choice either to regularly suspend its packet transfer for listening to the

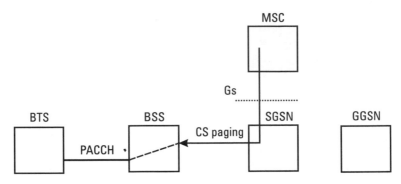

Figure 3.23 CS paging in packet transfer mode for network operation mode I.

CCCH paging channel or ignore the circuit-switched paging. The network may also send the circuit-switched paging on the PACCH; in this case the MS need not suspend its packet transfer for the paging decoding. Figure 3.24 illustrates a paging in idle mode when the network operates in mode II.

Mode III

In network operation mode III, the network sends circuit-switched paging messages on CCCH channels and packet-switched paging messages on PCCCH channels if they are present, or otherwise on CCCH channels. The Gs interface is not present.

In the case of a packet transfer in progress, the MS may receive a circuit-switched paging either on the CCCH paging channels or on PACCH. The MS behavior is the same as that defined in mode II during a packet transfer.

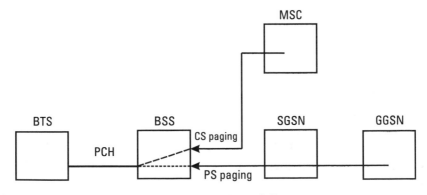

Figure 3.24 Paging in idle mode for network operation mode II.

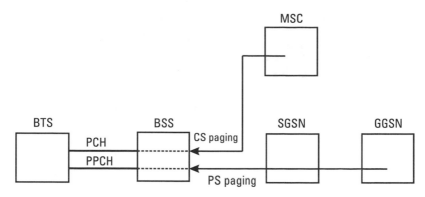

Figure 3.25 Paging in idle mode for network operation mode III.

In the case of a circuit-switched call in progress, the MS is not able to monitor PCCCH paging channels if they are present, or otherwise CCCH paging channels for a packet-switched paging. The MS is not authorized to suspend its circuit-switched call for listening PCHs. Figure 3.25 illustrates a paging in idle mode when the network operates in mode III.

3.5.3.2 GPRS Attach

In order to access GPRS services, an MS performs an IMSI attach for GPRS services to signal its presence to the network. During the attach procedure, the MS provides its identity, either a temporary identifier or *packet temporary mobile station identity* (P-TMSI) previously allocated by the SGSN, or an IMSI identifier when P-TMSI is not valid. When the MS is GPRS-attached, an MM context is established between the MS and the SGSN. This means that information related to this MS (i.e., IMSI, P-TMSI, cell identity, and RA) is stored in the SGSN. A GPRS-attached mobile is localized by the network at least at RA level and may be paged at any moment in GMM STANDBY state.

In network operation modes II or III, the MS both IMSI- and GPRS-attached is obliged to initiate separate attach procedures for circuit-switched and packet-switched services with MSC/VLR and SGSN entities.

In network operation mode I, there is a combined IMSI and GPRS attach procedure for MSs wishing to be configured in class A or in class B. In this case, the MS initiates an attach procedure for circuit-switched and packet-switched services with the SGSN. As the Gs interface is present between the MSC/VLR and SGSN, the latter is able to forward the IMSI attach request to the MSC/VLR.

3.5.3.3   GPRS Detach

A GPRS MS can no longer access GPRS service when it is GPRS-detached. The detach may be explicit or implicit. It is explicit when signaling is exchanged between the MS and the network. The detach is implicit when the network detaches the MS without any notification, and it may occur when the network does not detect any activity related to the MS for a certain amount of time.

An MS both IMSI- and GPRS-attached (MS configured in class A or class B) in a network that operates in network operation modes II or III is obliged to initiate separate procedures for IMSI detach and GPRS detach with MSC/VLR and SGSN entities.

An MS both IMSI- and GPRS-attached (MS configured in class A or class B) in a network that operates in network operation mode I may perform an IMSI and GPRS combined detach procedure. As Gs interface is present in mode I between the MSC/VLR and SGSN entities, the request of IMSI detach included in the request of combined detach is forwarded to the MSC/VLR entity by the SGSN entity.

3.5.3.4   Security Aspects

Principles of Authentication and Kc Key Establishment

The authentication procedure is equivalent to that existing in GSM, with the difference being that it is handled by the SGSN entity. This procedure allows the LLC layer between the MS and the SGSN to be protected against unauthorized GPRS calls. The GPRS authentication uses a nonpredictable number provided by HLR/AUC. From this random number, the MS and the HLR/AUC calculate a number called the SRES by using an algorithm A3 and a key Ki specific to the GPRS subscriber.

The establishment of ciphering key Kc for a GPRS subscriber is also performed during authentication procedure from the random number provided by HLR/AUC. The ciphering key Kc is also computed from the random number by using an algorithm A8 and a key Ki belonging to the GPRS subscriber. The principle of GPRS authentication is illustrated in Figure 3.26.

User Identity Confidentiality

The network guarantees user identity confidentiality while accessing GPRS radio resources. The user identity confidentiality is ensured by identifiers

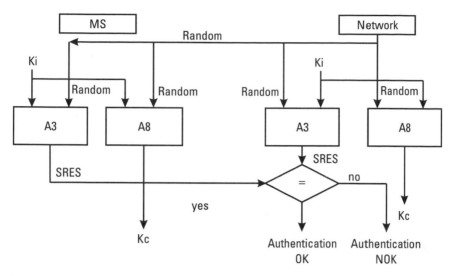

Figure 3.26 Principle of GPRS authentication.

such as the P-TMSI and *temporary logical link identity* (TLLI). These identifiers are solely known between the MS and the SGSN, and are used on the radio interface to identify the called or calling MS.

P-TMSI is locally allocated by the SGSN. As this attribute is handled locally at the SGSN level, the change of P-TMSI may take place at every SGSN change. The P-TMSI reallocation can be performed during the attach procedure, during the RA update procedure, or during the P-TMSI reallocation procedure requested by the network.

TLLI allows for identification of a GPRS subscriber. This identifier is deduced from the P-TMSI structure. The relation between TLLI and IMSI is only known by the MS and the SGSN. This identifier is calculated by the GMM entity and is sent to the RLC/MAC entity in the MS and the BSS. The SGSN may send a P-TMSI signature to the MS during the GPRS attach procedure or during the RA update procedure.

In this case, the MS must include this signature in the next attach procedure or in the next RA update procedure. The P-TMSI signature conveys the proof that the P-TMSI returned by the MS is the one allocated by the SGSN. The SGSN checks the signature sent by the MS and the one sent previously to the MS. If these values do not match, then the SGSN initiates a MS authentication procedure.

Identity Verification

The SGSN may ask for MS identity if the Gf interface is present between the SGSN and the EIR entity. That way, the SGSN may compare the IMEI identifier returned by the MS and the one saved in the EIR database.

Call Ciphering

The network provides the call confidentiality by ciphering it. The data ciphering for GPRS is performed at the LLC frame level between the MS and the SGSN, unlike in GSM, where the ciphering is performed on the radio interface between the MS and the BTS.

During GPRS ciphering, an XOR operation is performed between an LLC frame and a mask. This latter has the same length as the LLC frame. During the GPRS deciphering, an XOR operation is performed between a ciphered LLC frame and the same mask used for ciphering. The mask for the ciphering and deciphering procedures is calculated from ciphering algorithm A5 with the following input parameters:

- Key Kc, determined during authentication procedure;
- Direction, indicating the transmission direction (uplink or down-link);
- Input, which depends on the LLC frame type (acknowledged mode—incremented value for each new LLC frame; unacknowledged mode—value derived from the LLC header).

The ciphering and deciphering procedure is illustrated in Figure 3.27.

3.5.3.5 Location Updating Procedures

When a GPRS MS camps on a new cell, it reads the *cell identifier* (CI), the RAI, and the LAI. If at least one of these identifiers has changed, the MS may initiate one of the GPRS location procedures:

- Notification of cell update;
- Normal RA update;
- Combined RA and LA update.

If a GPRS-attached MS detects a new cell within its current RA, it performs a cell update procedure when it is in GMM READY state.

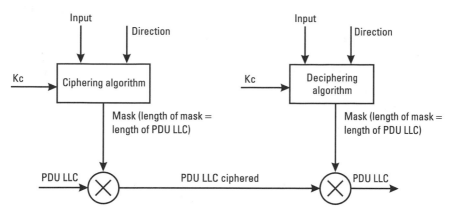

Figure 3.27 Ciphering and deciphering procedure.

If a GPRS-attached MS detects a new RA, it performs an RA update procedure. This procedure may also occur at the expiry of a periodic timer or at the end of a circuit-switched connection. In fact, during a circuit-switched connection, if the MS changes RA then the SGSN is not notified of this change.

If an MS that is both IMSI- and GPRS-attached has detected a new LA in a network that operates in network operation mode I, then it performs a combined RA and LA update procedure.

If an MS both IMSI- and GPRS-attached has detected a new LA in a network that operates in network operation modes II or III, then it first performs an LA update procedure, and then an RA update procedure. The LA update is performed by the MM entity and requires the establishment of a dedicated connection.

If a GPRS MS detects a new cell in a new RA, then it performs an RA update procedure. This procedure may also occur at expiry of a periodic timer or at the end of a dedicated connection.

3.6 PDP Context

A PDP context specifies access to an external packet-switching network. The data associated with the PDP context contains information such as the type of packet-switching network, the MS PDP address that is the IP address, the reference of GGSN, and the requested QoS. A PDP context is handled by the MS, SGSN, and GGSN and is identified by a mobile's PDP address within these entities. Several PDP contexts can be activated at the same time within a given MS.

A PDP context activation procedure is used to create a PDP context. This procedure may be initiated either by the MS or by the network. The MS is always GPRS-attached before PDP context negotiation. The PDP context activation may be performed:

- Automatically, if it is generated during a given procedure to perform a GPRS data transfer;
- Manually, if it is generated by user intervention.

A PDP context deactivation procedure is used to remove a PDP context. This procedure may be initiated either by the MS or by the network (SGSN or GGSN). The PDP context deactivation may be generated either during an application deactivation or during the GPRS detach or delete subscriber data procedures.

A PDP context modification procedure is used to modify the PDP context. This procedure may be initiated either by the MS or by the network in order to change QoS parameters or *traffic flow template* (TFT) parameters.

All PDP context procedures are handled by the SM protocol between the MS and the SGSN and by the GTP between the SGSN and the GGSN.

3.7   BSS Packet Flow Context Definition

A BSS Packet Flow Context contains the aggregate BSS QoS profile that is identical or similar for one or more activated PDP contexts. A BSS Packet Flow Context can be shared between several mobiles; several BSS Packet Flow Contexts can be defined. A BSS Packet FLow Context may be created, modified, or deleted every time a PDP context is activated, modified, or deleted.

A *packet flow identifier* (PFI), assigned by the SGSN, is used to identify each BSS Packet Flow Context. The PFI is assigned to the BSS at the creation of the BSS Packet Flow Context. It is assigned to the mobile when accepting the activation of the PDP context.

Three PFI values are reserved for best-effort service, SMS, and signaling.

Whenever the BSS receives an LLC PDU to transmit either in the direction of the SGSN or in the direction of the mobile, it deduces the QoS profile to use for the transmission from the PFI associated to the PDU.

The BSS Packet Flow contect concept has been introduced in the GPRS Release 99 recommendations.

3.8   Gb Interface

The Gb interface connects the BSS and the SGSN. It allows for the exchange of signaling information and user data. Many users are multiplexed on the same physical resource. Resources are allocated to the user only during activity periods; after these periods, resources are immediately released and reallocated to other users. This is in contrast to the GSM A interface where one user has the sole use of a dedicated physical resource during the lifetime of a call.

No dedicated physical resources are required to be allocated for signaling purposes. Signaling and user data are sent in the same transmission plane. Figure 3.28 shows the transmission plane on the Gb interface.

Transmission over the Gb interface is based on frame relay. *Point-to-point* (PTP) physical lines or an intermediate frame relay network can be used to connect the SGSN and the BSS.

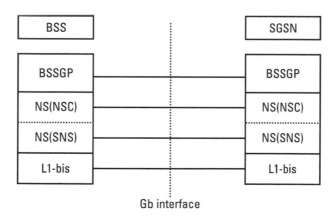

Figure 3.28   Transmission plane on Gb interface.

3.8.1   NS Layer

The NS layer provides a frame-based, multiplexed link layer transport mechanism across the Gb interface that relies on the frame relay protocol.

The NS layer has been split into two sublayers, *subnetwork service* (SNS) and *network service control* (NSC) in order to make one sublayer independent of the intermediate transmission network. SNS is based on frame relay but NSC is independent of the transmission network. Later, it will be possible to change the transmission network (e.g., with an IP network) without changing the NSC sublayer.

Peer-to-peer communication across the Gb interface between the two remote NS entities in the BSS and the SGSN is performed over virtual connections. The NS layer is responsible for the management of the virtual connections between the BSS and the SGSN (verification of the availability of the virtual connections, initialization, and restoring of a virtual connection). It provides information on the status and the availability of the virtual connections to the BSSGP layer. It ensures the distribution of upper-layer PDUs between the different possible virtual connections (load-sharing function).

SNS provides access to the intermediate transmission network (i.e., the frame relay network). NSC is responsible for upper-layer data (BSSGP PDUs) transmission, load sharing, and virtual connection management.

3.8.2   BSSGP Principle

The BSSGP layer ensures the transmission of upper-layer data (LLC PDUs) from the BSS to the SGSN or from the SGSN to the BSS. It ensures the transmission of GMM signaling and NM signaling.

The peer-to-peer communication across the Gb interface between the two remote BSSGP entities in the BSS and the SGSN is performed over virtual connections. There is one virtual connection per cell at BSSGP layer. Each virtual connection can be supported by several layer 2 links between the SGSN and the BSS.

The BSSGP layer is responsible for the management of the virtual connections between the SGSN and the BSS (verification of the availability of the virtual connections, initialization and restoring of a virtual connection). The BSSGP layer also ensures the data flow control between the SGSN and the BSS.

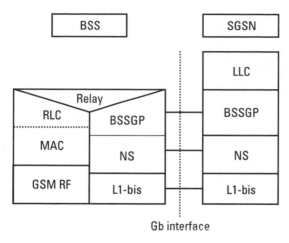

Figure 3.29 BSSGP position.

There is a one-to-one relationship between the BSSGP in the SGSN and in the BSS. That means if one SGSN handles several BSSs, the SGSN must have one BSSGP protocol machine for each BSS. Figure 3.29 shows the position of the BSSGP layer within the BSS and the SGSN.

3.9 GPRS Backbone Network Architecture

Figure 3.30 shows the architecture of the GPRS backbone network (see gray boxes) made up of GSNs.

All PDUs conveyed in the GPRS backbone network across the Gn/Gp interface are encapsulated by GTP. GTP allows IP PDUs to be tunneled through the GPRS backbone network and allows signaling exchange to be performed between GSNs. UDP/IP are backbone network protocols used for user data routing and control signaling.

3.9.1 Tunneling

A GTP tunnel is a two-way PTP path between two GSNs used to deliver packets between an external PDN and an MS. A GTP tunnel is created during a PDP context activation procedure. A GTP tunnel is identified in each GSN node by a *tunnel endpoint identifier* (TEID), a GSN IP address, and a

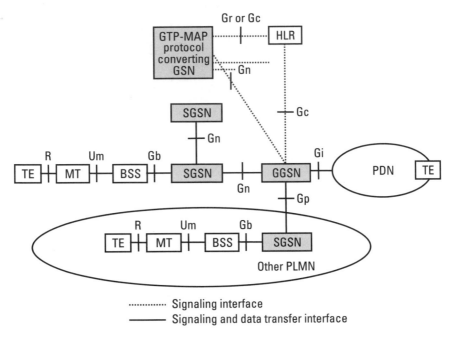

Figure 3.30 Architecture of GPRS backbone network. (*From:* [1].)

UDP port number. These identifiers are contained in IP and GTP PDU headers. There are two types of GTP tunnels:

- GTP-U tunnel (user plane), defined for each PDP context in the GSNs;
- GTP-C tunnel (control plane), defined for all PDP contexts with the same PDP address and *access point network* (APN).

The IP datagram tunneled in a GTP tunnel is called a T-PDU. With the tunneling mechanism, T-PDUs are multiplexed and demultiplexed by GTP between two GSNs by using the TEID field present in the GTP headers, which indicates the tunnel of a particular T-PDU. A GTP header is added to the T-PDU to constitute a G-PDU (or GTP-U PDU), which is sent in an UDP/IP path, a connectionless path between two endpoints.

Figure 3.31 illustrates a tunneling mechanism for IP packet sending toward MS.

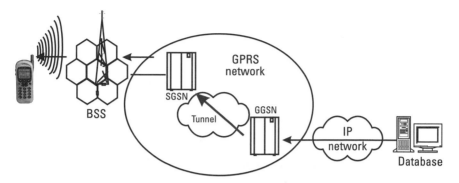

Figure 3.31 Tunneling mechanism for IP packet sending toward MS.

All signaling procedures (path management, tunnel management, location management, mobility management) between GSNs are tunneled in a GTP tunnel in the control plane. A GTP header is added to the GTP signaling message to constitute a GTP-C PDU, which is sent in a UDP/IP path.

3.9.2 Path Protocols

The UDP/IP path protocol is used to convey GTP signaling messages between GSNs or T-PDU in connectionless mode. Each UDP/IP path may multiplex several GTP tunnels. An endpoint of the UDP/IP path is defined by an IP address and a UDP port number. For UDP/IP path, the IP source address is the IP address of the source GSN, while the IP destination source is the IP address of the destination GSN. Note that the IP addresses of GSNs within the GPRS backbone network are private. This means that GSNs are not accessible from the public Internet.

Reference

[1]   3GPP TS 23.060 Service Description; Stage 2 (R99).

Selected Bibliography

Lagrange, X., P. Godlewski, and S. Tabbane, *Réseaux GSM-DCS: des principes à la norme*, Hermès, 2001.

3GPP TS 03.60 Service Description; Stage 2 (R97).

3GPP TS 03.64 Overall Description of the GPRS Radio Interface; Stage 2 (R99).

3GPP TS 23.007 Restoration Procedures (R99).

3GPP TS 24.008 Mobile Radio Interface Layer 3 Specification; Core Network Protocols—Stage 3 (R99).

3GPP TS 29.002 Mobile Application Part (MAP) Specification (R99).

3GPP TS 29.060 GPRS Tunnelling Protocol (GTP) across the Gn and Gp Interface.

3GPP TS 29.061 Interworking Between the Public Land Mobile Network (PLMN) Supporting Packet Based Services and Packet Data Networks (PDN) (R99).

4

Radio Interface: Physical Layer

GPRS uses the same underlying radio interface principles as GSM. The notions of time slot, frame, multiframe, hyperframe, and logical channels, described in Chapter 1, also apply to GPRS. All the participants involved in the standardization process, operators, handsets, and network vendors have taken care to minimize the impacts on the GSM radio interface. Although the main principles are upheld, new options have been introduced at the *physical link layer* (PLL) level during the specification phase of the GPRS standard, such as new logical channels, a new multiframe, and new coding schemes. New power control algorithms are specified for uplink and downlink to reduce the level of interference caused by the transmitters. A link adaptation mechanism is used to change the coding scheme according to the radio conditions in order to find the best trade-off between error protection and achieved throughput. We provided an overview of these additions in Section 3.3.1. Section 4.1 describes them in greater detail.

At the RF physical layer level, the main characteristic is the possibility of managing multislot configurations for GPRS MSs, in order to provide high-rate packet-switched services. New RF requirements have been introduced on the transmitter and on the receiver path. The details of the RF requirements are described in Section 4.2.

At the end of the chapter, case studies are given to show implementation constraints or issues concerning the GPRS system design.

4.1 PLL

This section describes the mechanisms that are used to guarantee an acceptable radio link quality between the MS and the BTS. After the discussion of physical channels, we will look at issues such as channel coding, uplink and downlink power control, and radio environment quality measurements.

4.1.1 Mapping of Logical Channels on the 52-Multiframe

The concept of logical and physical channels described for GSM also exists in GPRS. The difference with GSM is the possibility of dynamically configuring the mapping of the logical channels onto the physical channels. This feature allows the system to adapt itself to the network load, by allocating or releasing resources whenever needed.

The GPRS multiframe length is 52 TDMA frames, and it is organized into 12 blocks (B0 to B11) of 4 consecutive TDMA frames, plus 4 idle frames. A physical channel is referred to as a PDCH. A PDCH may be fully defined by ARFCN and a time slot pairing. Radio blocks, consisting of four bursts on a given PDCH, are used to convey the logical channels, transmitting either data or signaling (see Section 3.3.1.2 for the definition of the GPRS logical channels). The first and third idle frames within the 52-multiframe are used for the PTCCH on both uplink and downlink. This is represented in Figure 4.1.

A PDCH that supports PCCCHs is called a master channel because it carries all control signaling on PCCCH for packet transfer establishment. A

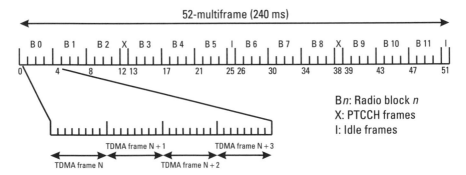

Figure 4.1 The 52-multiframe.

master channel also carries user data (PDTCH) and dedicated signaling (PACCH). The other PDCHs that do not support PCCCHs are called slave PDCHs. They carry only user data and dedicated signaling. The logical channels are dynamically mapped onto the multiframe. This is possible by means of different parameters that are transmitted to the MS in order to specify the number of blocks that are reserved for the different channels.

4.1.1.1   Uplink Channel Organization

A master channel configuration on the uplink may contain the following packet data logical channels: PRACH + PDTCH + PACCH + PTCCH. To map these channels on the multiframe, the MS uses an ordered list of blocks: B0, B6, B3, B9, B1, B7, B4, B10, B2, B8, B5, B11. A first group of blocks in this list is used for PRACH; a second group is used for PDTCH and PACCH. The PTCCH is not mapped dynamically, as seen in Figure 4.1 (TDMA frames 12 and 38).

The network notifies a PRACH occurrence on an uplink PDCH through the use of a specific USF field in the previous downlink block occurrence on the corresponding PDCH. Also, the network may define a fixed part of the 52-multiframe for PRACH use. In this case, the parameter BS_PRACH_BLKS (from 0 to 12), broadcast on the PBCCH, gives the uplink block occurrences that are reserved for the PRACH. The remaining blocks in the ordered list are used for PDTCHs and PACCHs. This is illustrated in Figure 4.2.

A slave configuration for the uplink can contain the following packet data logical channels: PDTCH + PACCH + PTCCH. A PDTCH (data) or PACCH (signaling) block may occur on any uplink radio block. The logical channel type is indicated by the *payload type* (PT) contained in the block header. The MS transmits PACCH blocks in radio occurrences indicated by a network polling message.

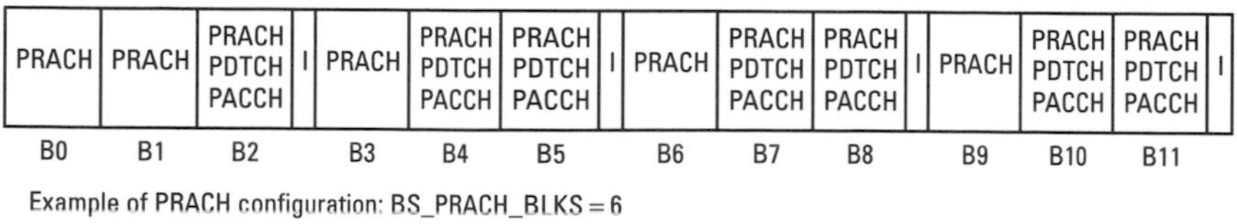

PRACH	PRACH	PRACH PDTCH PACCH	I	PRACH	PRACH PDTCH PACCH	PRACH PDTCH PACCH	I	PRACH	PRACH PDTCH PACCH	PRACH PDTCH PACCH	I	PRACH	PRACH PDTCH PACCH	PRACH PDTCH PACCH	I
B0	B1	B2		B3	B4	B5		B6	B7	B8		B9	B10	B11	

Example of PRACH configuration: BS_PRACH_BLKS = 6

Figure 4.2   Master channel configuration example on uplink.

4.1.1.2   Downlink Channel Organization

A master configuration for the downlink may contain one of the following packet data logical channel combinations:

- PBCCH + PCCCH + PDTCH + PACCH + PTCCH;
- PCCCH + PDTCH + PACCH + PTCCH, with
  PCCCH = PAGCH + PPCH.

As for the uplink master channels, the mapping of logical channels on the radio blocks is based on the ordered list B0, B6, B3, B9, B1, B7, B4, B10, B2, B8, B5, B11. The first block B0 is reserved for PBCCH. If more blocks are allocated for PBCCH (up to four radio blocks per 52-multiframe), then the PBCCH allocation follows the ordered list of blocks (B6, B3, and B9). The next radio blocks in the ordered list are reserved for PAGCH, and the remaining blocks are used for PPCH, PAGCH, PDTCH, and PACCH.

The BCCH gives information on the PDCH that carries PBCCH. Several parameters are then broadcast on the PBCCH to indicate the mapping of the master channels:

- BS_PBCCH_BLKS: number of blocks (1 to 4) reserved for the PBCCH within the 52-multiframe;
- BS_PAG_BLKS_RES: number of blocks (0 to 12) reserved for PCCCH within the 52-multiframe where PPCH and PBCCH are excluded; if a reserved occurrence is not used by a PAGCH block, then it may be used by a PDTCH or PACCH block.

Let us take the example of Figure 4.3 (first part), where BS_PBCCH_BLKS is equal to 4 and BS_PAG_BLK_RES is equal to 5. In this example, the PBCCH occurs in blocks B0, B6, B3, B9, and the PAGCHs, PACCHs, and PDTCHs are mapped on blocks B1, B7, B4, B10, B2. The remaining blocks are reserved for the PPCHs, PAGCHs, PDTCHs, and PACCHs on blocks B8, B5, and B11.

On master PDCHs that do not carry PBCCH, the first BS_PBCCH_BLKS blocks of the list are used for PDTCHs and PACCHs. The mapping of the other PPCHs, PAGCHs, PACCHs, and PDTCHs remains the same as on the master PDCH carrying PBCCH. The second part

PBCCH	PAGCH PDTCH PACCH	PAGCH PDTCH PACCH	I	PBCCH	PAGCH PDTCH PACCH	PPCH	I	PBCCH	PAGCH PDTCH PACCH	PPCH	I	PBCCH	PAGCH PDTCH PACCH	PPCH	I
B0	B1	B2		B3	B4	B5		B6	B7	B8		B9	B10	B11	

Master channel supporting both PCCCH and PBCCH, with BS_PBCCH_RES = 4 and BS_PAG_BLKS_RES = 5

PDTCH PACCH	PAGCH PDTCH PACCH	PAGCH PDTCH PACCH	I	PDTCH PACCH	PAGCH PDTCH PACCH	PPCH	I	PDTCH PACCH	PAGCH PDTCH PACCH	PPCH	I	PDTCH PACCH	PAGCH PDTCH PACCH	PPCH	I
B0	B1	B2		B3	B4	B5		B6	B7	B8		B9	B10	B11	

Master channel supporting PCCCH but not PBCCH, with BS_PBCCH_RES = 4 and BS_PAG_BLKS_RES = 5

Figure 4.3   Master channel configuration example on downlink.

of Figure 4.3 shows an example of this configuration with BS_PBCCH_RES and BS_PAG_BLKS_RES parameters equal respectively to 4 and 5.

A downlink slave configuration contains the following packet data logical channel combination: PDTCH + PACCH + PTCCH. A PDTCH or PACCH block may occur on any radio block. Data transfer is carried on PDTCH while signaling is carried on PACCH.

4.1.2   Channel Coding

4.1.2.1   PDTCH

Four coding schemes, CS-1 to CS-4, are used for the GPRS PDTCHs. They offer different levels of protection, and the CS to be used is chosen by the network according to the radio environment. If, for instance, the *carrier to interference ratio* (C/I) is high, low protection is applied, to achieve a higher data rate, and if the C/I is low, the level of protection is increased, which leads to a lower data rate. This mechanism is called link adaptation.

Coding schemes CS-1 to CS-4 are mandatory for MSs supporting GPRS, but with the exception of CS-1 they may not be supported by the network. Figures 4.4 and 4.5 show the coding principle of one radio block for CS-1 to CS-3 and CS-4, respectively. On the bottom parts of the figures are represented the four NBs on which the block is mapped; each one is transmitted in one TDMA frame.

The first step of the coding procedure consists of adding a *block check sequence* (BCS) coded on 16 bits for error detection. For CS-2, CS-3, and CS-4, the next step is the precoding of the USF field. The 3-bit field is

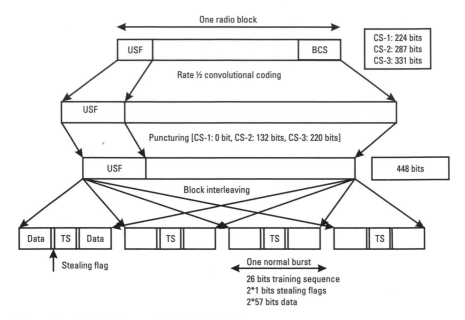

Figure 4.4   Radio block encoding for CS-1 to CS-3.

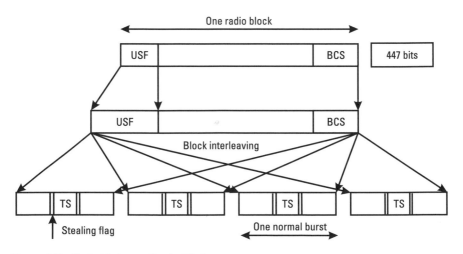

Figure 4.5   Radio block encoding for CS-4.

Table 4.1

Convolutional Code Parameters for CS-1, CS-2, and CS-3

Coding Schemes	Polynomials	Code Rate
CS-1, CS-2, and CS-3	$G0 = D^4 + D^3 + 1$	1/2
	$G1 = D^4 + D^3 + D + 1$	

mapped onto a 6-bit word in CS-2 and CS-3, and on a 12-bit word in CS-4. The USF is used as a token for uplink multiplexing. It needs to be decoded with a very low error rate, by the different mobiles allocated on the same PDCH. That is why its coding is more robust than the rest of the block.

For CS-1, CS-2, and CS-3, the next step involves adding four tail bits prior to a rate 1/2 convolutional encoding stage for error correction. This code is punctured to give the desired coding rate. The same convolutional code is used for the three coding schemes (see Table 4.1) but with a different puncturing rate. The interest of the puncturing is to allow different coding rates with the same convolutional code. It consists on not transmitting some of the bits that are output by the encoder. At the reception, the hole positions are known by the receiver. They are filled with zero before the decoding. For CS-4, there is no convolutional coding. Note that after the convolutional encoding stage, the 3-bit USF field is mapped on a 12-bit word in CS-2, CS-3 (as it is in CS-4), and on a 6-bit word in CS-1.

The coded data obtained (456 bits) is then interleaved over four bursts. The burst format is the NB, as described in Chapter 1.

In GSM, along with each block of 456 coded bits there is a *stealing flag* (SF) pattern, indicating whether the block belongs to the TCH or to the FACCH. As the FACCH is not used in GPRS, the SFs, at the extremities of the 26-bit TS code, are used to signal the coding scheme used for the block. With the SFs (sequence of 8 bits in a radio block, with 2 bits in each burst), the receiver is able to detect which coding scheme is used for that block, with the correspondence as shown in Table 4.2. Note that the MS always transmits with a CS commanded by the network, whereas in downlink it performs an SF detection prior to the decoding of the block.

Table 4.3 summarizes the characteristics of the different coding schemes. CS-1 uses the same coding scheme as specified for SACCH in GSM. The data rate includes the RLC/MAC header and RLC information.

Table 4.2

Mapping Between Coding Scheme and Stealing Flags

Coding Scheme	Stealing Flag Pattern
CS-1	11111111
CS-2	11001000
CS-3	00100001
CS-4	00010110

Link Adaptation

The basic principle of link adaptation is to change the coding scheme according to the radio conditions. When the radio conditions get degraded, a more protected CS (more redundancy by reduction of the overall code rate) is chosen, leading to a lower throughput. In contrast, when the radio conditions become better, a less protected CS is chosen, leading to a higher throughput. The radio conditions are determined by the network based on the measurements performed by the MS and the BTS.

The way the "used CS" is signaled by the transmitter to the receiver (in-band signaling by the SFs, as described above) allows the transmitter to adapt its transmission rate without notifying the receiver.

Depending on the radio conditions (Doppler shift due to the mobile speed, multipath, interference) in which the MS operates, the network modifies its transmission rate (CS change) in downlink and orders the use of another CS to the mobile in uplink. The ordered CS for use in uplink is sent within RLC/MAC control messages on the PACCH.

Table 4.3

Coding Parameters for GPRS Coding Schemes

Scheme	Code Rate	USF	Precoded USF	Radio Block Excluding USF and BCS	BCS	Tail Bits	Coded Bits	Punctured Bits	Data Rate (Kbps)
CS-1	1/2	3	3	181	40	4	456	0	9.05
CS-2	≈2/3	3	6	268	16	4	588	132	13.4
CS-3	≈3/4	3	6	312	16	4	676	220	15.6
CS-4	1	3	12	428	16	—	456	—	21.4

The network evaluates the radio conditions based on quality and received level measurements either performed by the mobile and reported to the network (downlink transfer) or performed by the network (uplink transfer).

4.1.2.2 PCCCH Channel Coding

For all GPRS packet control channels other than PRACH and PTCCH on uplink, the coding scheme CS-1 is always used.

For ABs on PRACH, two coding schemes are defined:

- Packet AB that carries 8 information bits, the coding of which is the same as for GSM;

- Extended packet AB that carries 11 information bits. The coding is the same as for the normal AB except that puncturing is used to increase the information field (which slightly decreases the performance).

The extended AB carries more information, which allows the mobile to give more information when requesting the establishment of a TBF. The use of normal AB or extended AB is controlled by the network parameter ACCESS_BURST_TYPE broadcast on BCCH or PBCCH.

4.1.2.3 PTCCH Channel Coding

The PTCCH downlink channel coding is the same as for the GSM SACCH. In uplink, either packet ABs or extended packet ABs are sent depending on the network parameter ACCESS_BURST_TYPE. The information field is set to a fixed pattern.

4.1.3 Power Control

The power control mechanism allows for the adjustment of the radio transmitter output power in order to minimize the interference caused by the transmitter, while maintaining a good radio link quality. In GPRS, power control is used on the uplink and also on the downlink.

4.1.3.1 Uplink Power Control

The uplink power control can be performed in two ways. Either it is managed autonomously by the mobile, in which case the procedure is called open-loop power control, or it is partly controlled by the network, in which case it is called closed-loop power control.

Open-loop uplink power control relies on received power measurements (RXLEV) performed by the mobile on its serving cell. For this mode it is assumed that the path loss is the same in uplink and in downlink. The output power is therefore derived from the RXLEV: the higher the RXLEV, the lower the transmit power.

In closed-loop power control, the network can control either fully or partially the output power that is used by the mobile. The power control in this case can rely on RXLEV measurements performed by the MS on the downlink path or RXLEV and RXQUAL measurements performed by the network on the uplink path, or both types of measurements.

The MS calculates the RF output power value, P_{ch}, to be used on each individual uplink PDCH assigned to the MS with the following equation:

$$P_{ch} = \min\left[\left(\Gamma_0 - \Gamma_{ch} - \alpha * (C + 48) \right), P_{max} \right] \tag{4.1}$$

where Γ_{ch} is a parameter specific to the MS and to the PDCH, given to the MS in an RLC control message, Γ_0 is a constant parameter that is frequency band dependent, α is a parameter in the range [0,1], P_{max} is the maximum output power allowed in the cell, and C is the average received signal level. P_{max} is broadcast by the network; Γ_{ch} and α are parameters that are sent to the MS in dedicated signaling messages.

Equation (4.1) can be rewritten in the following way:

$$P_{ch} = \beta - \alpha * C \tag{4.2}$$

with P_{ch} limited to P_{max}.

Open-Loop Uplink Power Control

The open loop corresponds to the case where α is set to 1 and β has a constant value; α and β are set during the TBF establishment and are not modified during the TBF. The MS output power is directly deduced from the average received signal level C. Figure 4.6 gives an example of the variation of the output power during time depending on the average received signal level. Note that the output power is limited to P_{max}.

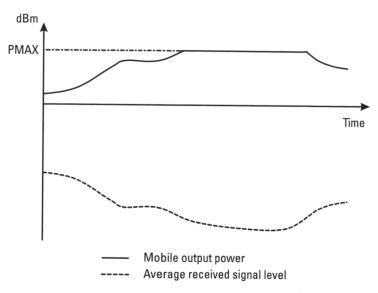

Figure 4.6   Example of uplink power control in open loop.

Closed-Loop Uplink Power Control

As described above, in closed-loop mode, the network can control the computed mobile output power. This control is performed through the sending of the parameters Γ_{ch} and α during the packet transfer. In case of pure closed-loop mode, the parameter α is set to 0 by the network and the MS output power is directly controlled by the network with the Γ_{ch} parameter (4.3). In this case the network performs uplink measurements (RXLEV, RXQUAL) and deduces the output power that must be applied by the mobile. One Γ_{ch} value is sent to the MS for each PDCH to control the computed power.

$$P_{ch} = \beta \text{ if } \alpha = 0 \tag{4.3}$$

More complicated algorithms that control the output power with both the Γ_{ch} and α parameters can be implemented in the network.

Received Signal Level Averaging Process

In packet transfer mode, the mobile may perform its measurements on the BCCH or on one allocated PDCH. The PC_MEAS_CHAN parameter broadcast by the network indicates on which type of channel the mobile must perform its measurements. When downlink power control is used by the BTS, the measurements must be performed on the BCCH.

The received signal level is averaged through a recursive filtering that is given by (4.4).

$$C_n = (1-F)*C_{n-1} + F*SS_n \tag{4.4}$$

C_n represents the average received signal level at the nth iteration, F is the forgetting factor of the recursive filter, and SS_n is either the received signal level averaged over the four bursts that compose the block when the measurements are performed on paging blocks or PDCH, or the received signal level of the measurement sample when the measurements are performed on the BCCH carrier.

F is a parameter that is broadcast by the network. Its value can be different depending on the RR state of the mobile and on the channel on which the measurements are performed. A summary of the parameters and their use is presented in Table 4.4.

Table 4.4
Summary of Uplink Power Control Parameters

Parameters	Transmission Mode of Parameters	Usage
P_{max}	Broadcast on BCCH or PBCCH	Maximum output power allowed in the cell
Γ_{ch}	In dedicated signaling	Closed-loop control parameter different on PDCH basis
α	Broadcast on BCCH or PBCCH and in dedicated signaling	Used to control trade-off between open- and closed-loop power control
T_{AVG_T}	Broadcast on BCCH or PBCCH	Used to compute forgetting factor F in packet transfer mode.
PC_MEAS_CHAN	Broadcast on BCCH or PBCCH	Used to order measurement on the BCCH or PDCH during packet transfer mode

Table 4.5
BTS Output Power Relative to Logical Channel Type

Logical Channel Type	Downlink Power Control Used	Presence of PBCCH	BTS Output Power
PBCCH/PCCCH	—	—	BCCH output power – Pb
PTCCH	—	Yes	BCCH output power – Pb
	—	No	BCCH output power
PDTCH	Yes	—	Any power level that respects the rules described later in section
	No	Yes	BCCH output power – Pb or no transmission
		No	BCCH output power or no transmission

4.1.3.2   Downlink Power Control

Downlink power control is optional on the BTS side. The operations are different according to the logical channel (see Table 4.5):

- The BTS must use constant output power on the PDCH that carries PBCCH or PCCCH. The power level used on those PDCHs can be lower than the one used on BCCH. This power level difference is broadcast by the network (Pb parameter).
- The output power on the PTCCH downlink is constant and is the same as on PBCCH or BCCH if PBCCH does not exist in the cell.
- On the other PDCH radio blocks, downlink power control can be used. This mechanism is based on the channel quality reports send by the mobile. The network must use the same output power for the four bursts that compose one radio block.

When downlink power control is used, the BSS must ensure that the output power is sufficient for the MS for which the RLC block is intended as well as the MS for which the USF is intended when operating in dynamic or extended dynamic allocation. Indeed, the RLC block can be addressed to a given mobile, but with a USF field intended for another, as shown in Figure 4.7.

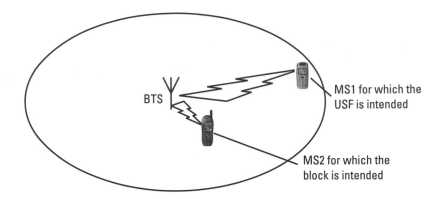

Figure 4.7   Downlink power control for a block addressed to two different mobiles.

Two modes of downlink power control have been defined in order to limit the variations on the BTS transmit power and to allow operation with standard implementation of *automatic gain control* (AGC) in the mobile (see Section 4.3.5.3).

- Power control mode A can be used with any allocation scheme;
- Power control mode B is intended for fixed allocation only.

The BTS power control mode is indicated during the TBF resource assignment by the parameter BTS_PWR_CTRL_MODE.

Both modes use the parameter P0, defined as a *power reduction* (PR) relative to BCCH output power. This parameter is included in the assignment message. BCCH level – P0 corresponds to the initial BTS output power when the network starts the downlink transfer. The mobile is therefore able to evaluate roughly the received level at the beginning of the transfer from the received level on BCCH. It uses this estimate for the setting of its AGC at the start of the transmission.

Power Control Mode A

In BTS power control mode A, the BTS limits its output power on blocks addressed to a particular MS to levels between (BCCH level – P0) and (BCCH level – P0 –10 dB). The BTS can change its output power from time slot to time slot within the same TDMA frame. The power level of blocks not addressed to the mobile does not have to exceed BCCH level – P0.

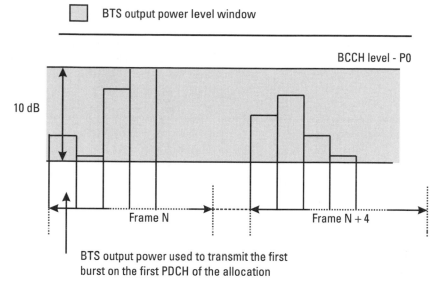

Figure 4.8 Example of downlink power control mode A.

Figure 4.8 shows an example of BTS output power limitations. In this example, the first four *time slots* (TSs) are allocated to a multislot mobile.

Power Control Mode B

In power control mode B, the BTS may use its full output power range (i.e., 30 dB). It uses the same output power on all blocks addressed to a particular MS within a TDMA frame. The BTS changes the output power no faster than one nominal power control step (2 dB) every 60 ms.

Moreover, when the BTS addresses a block to a particular mobile M1 and the block that is transmitted on the previous time slot is addressed to another mobile M2, if the block that is addressed to M2 is transmitted on a PDCH that is part of the M1 allocation, its transmit output power shall not exceed the transmit output power of the block that is addressed to M1 by more than 10 dB.

This is summarized in Figure 4.9, where the first four bursts of the TDMA frame (one TDMA frame = 8 TS) are represented, because they correspond in the example to the PDCH allocation of a given mobile. The block that is represented in gray on the figure is dedicated to another MS, and can therefore be sent with a different level, that can be up to 10 dB higher, as explained above.

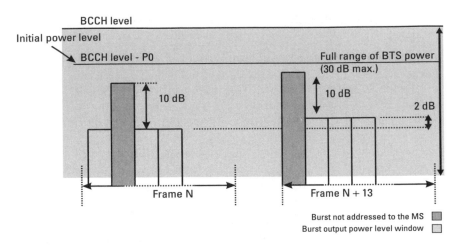

Figure 4.9   Example of downlink power control mode B.

PR Management

When downlink power control is used, the BTS may indicate, for some PDCH downlink blocks, the output power level used to send these blocks. This level is indicated by the PR field if present in the block. It gives the power level reduction used to send this block relative to BCCH level—P0.

Two PR management modes have been introduced: PR mode A and PR mode B. The PR mode is indicated during the assignment phase with the PR_MODE parameter. In PR mode A, the PR field of a block is calculated based on the BTS output power level in the direction of the addressed (RLC information) MS. This mode is suited for the case of adaptive antennas (the transmitted power is not the same in the different directions, and therefore an MS for which the block is not addressed cannot use the PR field). In PR mode B, for each block sent on a given PDCH, the BTS uses the same output power level for all the MSs with TBF on this PDCH. Since the PR field can be interpreted by all the mobiles, this mode cannot be used with adaptive antennas.

The network is not allowed to change the PR mode during a TBF. The information fields PR, P0, PR_mode, and BTS_PWR_CTRL_MODE may be used by the MS for its AGC algorithm.

Table 4.6 lists the parameters that are involved in the downlink power control procedure.

Table 4.6
Downlink Power Control Parameters

Parameters	Transmission Mode of Parameters	Usage
BTS_PWR_CTRL_MODE	During resource assignment	BTS downlink power control mode used on network side
PR_MODE	During resource assignment	PR mode associated with power control
P0	During resource assignment	Used to indicate the initial output power of the BTS
PR	RLC/MAC block	Power level reduction used to send the radio block
Pb	Broadcast on BCCH or PBCCH	Power level reduction of PBCCH and PCCCH compared to BCCH

4.1.4 MS Measurements

The measurements performed by the MS are described in Section 3.3.1.5. The network may request the mobile to perform measurements in packet idle mode for reasons other than reselection. These measurements are used by the network operator for frequency planning, and may be of two types:

- Power measurements on an additional frequency list;
- Interference measurements on an additional frequency.

4.1.4.1 Extended Measurements

The network broadcasts a list of frequencies to be measured for extended measurements. A reporting period is indicated by the network to describe the periodicity of the MS reports.

4.1.4.2 Interference Measurements

The network may request the mobile to perform interference measurements. These measurements involve estimating the received level on a PDCH that is not necessarily a broadcast channel. They are performed on different channels depending on the RR state:

- In packet idle mode—on a list of carriers broadcast by the network;
- In packet transfer mode—on the same carrier as the assigned PDCH.

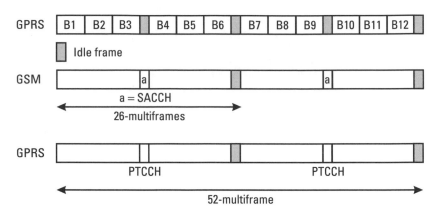

Figure 4.10 Collision between idle frames and SACCH frames or PTCCH frames.

The MS measures the received signal level during its idle frames, on a carrier allocated in its serving cell, to estimate the interference level due to neighbor cells. It may happen that the measurement is performed when a signal is transmitted in the MS cell, leading to a wrong estimation of the interference level due to neighbor cells. For instance, this signal may be a GSM SACCH block (as the structure of a 52-multiframe is the juxtaposition of two 26-multiframes) or a GPRS PTCCH occurrence—if the TA update procedure is used on this channel (see Figure 4.10).

The MS performs at least one interference measurement on a number of TSs, which is equal to the maximum number of receive TSs according to its multislot class. An instantaneous interference value is obtained for each time slot. In order to decorrelate the estimation from the signaling blocks (SACCH and PTCCH) from its own cell, it performs the measurement on at least two consecutive idle frames. Only the lowest measured level is kept, so that the signal measurement on the serving cell can be discarded. The received signal level samples $SS_{ch,n}$ on the PDCH are then averaged with a running average filter:

$$\gamma_{ch,n} = (1-d) * \gamma_{ch,n-1} + d * SS_{ch,n} \qquad (4.5)$$

where $\gamma_{ch,n}$ is the measured interference level, d is the forgetting factor, and n is the iteration index.

The interference measurements are not reported in packet idle mode but are reported in packet transfer mode in case of network request. For each time slot where interference measurements are performed, a running average $\gamma_{ch,n}$ value is reported.

4.2 RF Physical Layer

This section first describes the concept of multislot classes, which allows for transmission in several time slots in the TDMA frame. It then focuses on the transmitter and receiver requirements. For the former, the constraints introduced by multislot transmissions are presented. For the latter, the sensitivity and interference performance specifications are discussed, as well as GSM circuit-switched requirements such as blockings, intermodulation, and AM suppression, which also apply in GPRS Release 99. The transmitter and receiver requirements are defined in [1].

4.2.1 Multislot Classes

One of the most important GPRS characteristics on the air interface is the possibility of increasing the achievable bit rate by grouping together several channels. In order to do so with a reasonable impact on the MS in terms of implementation complexity, it has been decided to allow the RF transmission on several slots of a TDMA frame, with a number of restrictions, as listed below.

- Several bursts can be transmitted within a TDMA frame, but should all be on the same ARFC number (ARFCN; i.e., the same carrier frequency).

- Depending on the mobile capability, delay constraints are needed between the transmission and reception of bursts, and between reception and transmission, to allow the mobile to perform the adjacent cell measurements, or monitoring, as discussed in Chapter 1.

- If there are m time slots allocated to an MS for reception and n time slots allocated for transmission, the system requires that min (m,n) reception and transmission time slots have the same time slot number (TN; see Section 1.5.1 for the definition) within the TDMA frame.

Two types of MSs are defined: type 1 mobiles are not able to transmit and to receive at the same time, while type 2 mobiles are. For these two types, there exist different classes, depending on the capability of the MS in terms of complexity. The classes are called multislot classes since they refer to the ability of the mobile to support a communication on several time slots of the TDMA frame. For a given multislot class, the mobile is able to transmit on a maximum of Tx time slots, and to receive on a maximum of Rx time slots within a TDMA frame, but the sum Tx + Rx is limited. This means that the maximum of Tx slots and the maximum of Rx slots are not active at the same time.

The definition of the type 1 multislot classes relies on the following time constraints:

- T_{ta} is the maximum number of time slots allowed to the MS to measure an adjacent cell received signal and to get ready to transmit. This parameter is therefore used to set the minimum allowed delay between the end of a transmit or receive time slot and the next transmit time slot, with an adjacent cell measurement to be performed in between.

- T_{rb} relates to the number of TS needed by the MS, prior to a receive time slot, when no adjacent cell measurement is performed. It is the minimum delay between the end of a transmit or receive TS and the first next receive TS.

- T_{ra} is the minimum allowed delay in number of TS, between the end of a transmit or receive time slot and the next receive time slot, when an adjacent cell measurement is to be performed in between.

- T_{tb} is the minimum number of TS between the end of a receive or transmit TS and the first next transmit TS, without adjacent cell measurement in between.

These constraints have been chosen to give the mobile enough time for the frequency change between a receive or transmit slot, and the next receive or transmit slot. It also allows some time to measure an adjacent cell received signal level, which requires a measurement window (the window size is usually in the order of one time slot or less) and two frequency changes (from the transmit or receive frequency to the adjacent cell beacon frequency, and then back to the next receive or transmit time slot). Note that the time constraints T_{ta} and T_{tb} may be reduced by the amount of the TA, to derive the allowed time duration before transmission.

The existing multislot classes are given in Tables 4.7 (type 1 MSs) and 4.8 (type 2 MSs), with the Tx and Rx parameters defining the maximum

Table 4.7
Type 1 MS Multislot Classes

Type 1 MSs							
	Maximum Number of Slots			Minimum Number of Slots			
Multislot Class	Rx	Tx	Sum Rx+Tx	T_{ta}	T_{tb}	T_{ra}	T_{rb}
1	1	1	2	3	2	4	2
2	2	1	3	3	2	3	1
3	2	2	3	3	2	3	1
4	3	1	4	3	1	3	1
5	2	2	4	3	1	3	1
6	3	2	4	3	1	3	1
7	3	3	4	3	1	3	1
8	4	1	5	3	1	2	1
9	3	2	5	3	1	2	1
10	4	2	5	3	1	2	1
11	4	3	5	3	1	2	1
12	4	4	5	2	1	2	1
19	6	2	N/A	3	a)	2	b)
20	6	3	N/A	3	a)	2	b)
21	6	4	N/A	3	a)	2	b)
22	6	4	N/A	2	a)	2	b)
23	6	6	N/A	2	a)	2	b)
24	8	2	N/A	3	a)	2	b)
25	8	3	N/A	3	a)	2	b)
26	8	4	N/A	3	a)	2	b)
27	8	4	N/A	2	a)	2	b)
28	8	6	N/A	2	a)	2	b)
29	8	8	N/A	2	a)	2	b)

a) = 1 with frequency hopping or change from Rx to Tx;
= 0 without frequency hopping and no change from Rx to Tx;
b) = 1 with frequency hopping or change from Tx to Rx;
= 0 without frequency hopping and no change from Tx to Rx;
N/A: not applicable.

Table 4.8
Type 2 MS Multislot Classes

	Type 2 MS						
	Maximum Number of Slots			Minimum Number of Slots			
Multislot Class	Rx	Tx	Sum Rx+Tx	T_{ta}	T_{tb}	T_{ra}	T_{rb}
13	3	3	N/A	N/A	(*)	3	(*)
14	4	4	N/A	N/A	(*)	3	(*)
15	5	5	N/A	N/A	(*)	3	(*)
16	6	6	N/A	N/A	(*)	2	(*)
17	7	7	N/A	N/A	(*)	1	0
18	8	8	N/A	N/A	0	0	0

(*) 1 with FH, 0 without FH; N/A: not applicable.

uplink and downlink slots of the TDMA frame, the sum being the maximum value for Rx + Tx. The time constraints T_{ta}, T_{rb}, T_{ra}, T_{tb} corresponding to each class are also given in these tables.

Note that only one monitoring window (i.e., an adjacent cell power measurement window) is needed in a TDMA frame, so that only a couple (T_{ra}, T_{tb}) or (T_{ta}, T_{rb}) is needed to define a valid configuration of a given multislot class. Figure 4.11 further illustrates the different possible configurations allowed for a class 12 mobile.

One can notice that for classes 1 to 12 the sum of Tx + Rx slots, added to $T_{ra} + T_{tb}$, is always less than or equal to 8. This comes from the fact that the mobile, for these classes, is not full duplex capable (i.e., it is not able to transmit and to receive at the same time). The total sum of these constraints cannot be more than 8 time slots, which is the duration of a TDMA frame.

For classes 13 to 18, this constraint is not in use, since the mobile has either the ability to receive and transmit at the same time, or to receive or transmit and perform an adjacent cell measurement at the same time.

4.2.2   Transmitter Path Characteristics

The GPRS standard was designed to minimize the changes on the RF layer of standard GSM equipment. On the transmitter part, for both the MS and

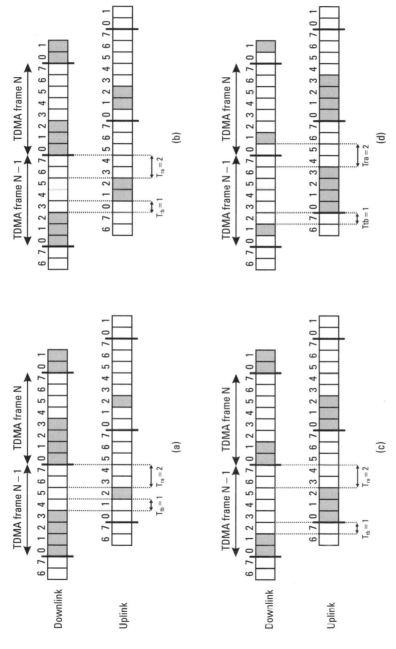

Figure 4.11 Allowed configurations for multislot class 12. (a) 4 RX + 1 TX, (b) 3 RX + 2 TX, (c) 2 RX + 3 TX, and (d) 1 RX + 4 TX.

BTS, it has therefore been decided to keep the existing power classes, as well as the power control ranges and steps. The changes due to GPRS on Tx are limited to the constraint of the multislot transmission, with a new power ramping template. Indeed, as seen in Section 4.1.3, power control is used on uplink, and independent transmit power on adjacent time slots, and therefore any combination of power control steps may be applied on several time slots of the same TDMA frame. The single slot power versus time mask is shown in Chapter 1, in Figure 1.11, for the NB.

For multislot transmissions, the constraint is slightly modified, as follows:

- The same template as for the single slot case is to be respected during the useful part of each burst and at the beginning and the end of the series of consecutive bursts.

- The output power during the guard period between every two consecutive active time slots cannot be greater than the level allowed for the useful part of the first time slot, or the level allowed for the useful part of the second time slot plus 3 dB, whichever is the highest.

- Between the active bursts, the output power must be lowered to a specified limit.

Figure 4.12 shows the power-versus-time-mask example for a two-slot transmitter, for different power configurations.

4.2.3   Receiver Path Characteristics

4.2.3.1   Reference Sensitivity Performance

In the case of GSM circuit-switched services (see Section 1.5.6.2), a minimum input level at the receiver is defined, at which the MS and the BTS are required to reach a certain level of performance.

For GSM voice service, for instance, we have seen that GSM900 small MSs (see definition in Section 1.5.6.1) are required to reach a given BER, FER, and RBER, for an *input signal level* (ISL) of −102 dBm. The performances (BER, FER, RBER) are different depending upon the channel profile (static, TU50, RA250, HT100), but the level is −102 dBm for all profiles.

The performance to be met in GPRS is the *block error rate* (BLER), referring to all erroneously decoded data blocks including any headers, SFs, data, and parity bits. Once a radio block, comprising four bursts, is received on a PDTCH, the mobile performs a de-interleaving and a decoding of the convolutional code, for coding schemes CS-1 to CS-3 (no convolutional

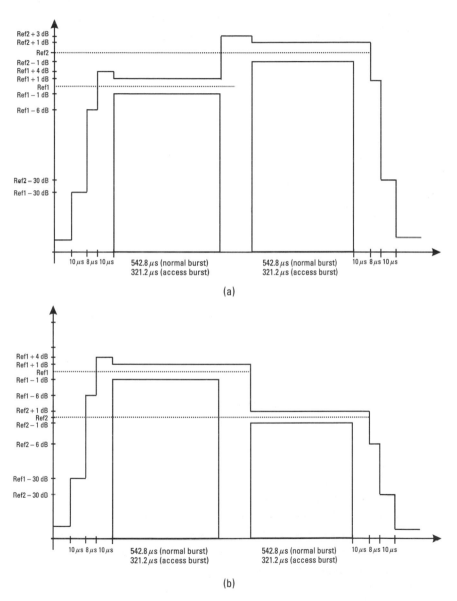

Figure 4.12 Multislot power versus time mask for the NB and for the AB: (a) power level is higher on first time slot; and (b) power level is higher on second time slot.

code is used for the coding scheme CS-4). The parity bits from the block code are then calculated, based on the received bits. If the calculated parity sequence is different from the received sequence, the block is declared erroneous. The ratio of these erroneous blocks to the number of received blocks is therefore an estimation of the BLER. The BLER is also defined for the USF and it refers to the ratio of erroneously decoded USF words over the number of received blocks.

The principle of the sensitivity performance definition is somewhat different in GPRS than in GSM circuit, in the sense that the BLER level to be reached for the MS and the BTS is not dependent on the channel profile. Indeed, for all the channel profiles:

- A BLER of 10% will be achieved on the PDTCHs, for all the coding schemes CS-1 to CS-4.
- A BLER of 1% will be achieved on the USF field.
- A BLER of 15% will be achieved on the PRACHs.

These requirements are valid for both the BTS and MS, and for different propagation conditions. The idea behind this is to ensure a constant quality of the data link (constant BLER) in all the propagation conditions. Nevertheless, the level for which these performance levels are to be reached are different according to the equipment type (MS or BTS), the channel profile, and the coding scheme.

Table 4.9 gives the ISL for the sensitivity performance, for the different coding schemes and propagation profiles, for a GSM-900 or GSM-850 normal BTS. The figures are slightly different for DCS-1800 and PCS-1900. For the MS, all the ISLs for the sensitivity case can be derived from the BTS performance, by adding an offset of 0, +2, or +4 dB (according to the MS power class and its band of operation) to the ISL figures.

Note that the specification for PDTCH/CS-1 applies also for PACCH, PBCCH, PAGCH, PPCH, and PTCCH/D.

To avoid the blinding of the receiver, the sensitivity specification is not required if the received level on either the time slot immediately before or immediately after a received time slot is greater than the desired time slot level by more than 50 dB for the BTS, or 20 dB for the MS. Also, for the MS, these specifications are not to be fulfilled if the received level on any of the time slots belonging to the multislot configuration is greater than the desired (that is the one on which the BLER is actually measured) time slot level by more than 6 dB.

Table 4.9
ISL at Reference Performance for Normal BTS

	GSM-900 and GSM-850 Normal BTS				
		Propagation Conditions			
Type of Channel	**Static**	**TU50 (No FH)**	**TU50 (Ideal FH)**	**RA250 (No FH)**	**HT100 (No FH)**
PDTCH/CS-1 (dBm)	−104	−104	−104	−104	−103
PDTCH/CS-2 (dBm)	−104	−100	−101	−101	−99
PDTCH/CS-3 (dBm)	−104	−98	−99	−98	−96
PDTCH/CS-4 (dBm)	−101	−90	−90	No requirement	
USF/CS-1 (dBm)	−104	−101	−103	−103	−101
USF/CS-2 to 4 (dBm)	−104	−103	−104	−104	−104
PRACH/11 bits (dBm)	−104	−104	−104	−103	−103
PRACH/8 bits (dBm)	−104	−104	−104	−103	−103

4.2.3.2 Interference Performance

The same BLER performance (10% for PDTCH, 15% for PRACH, and 1% for the USF) is required when interference is added to the wanted signal. The interference specifications are defined for a desired signal input level depending on the channel profile, and for a random, continuous, GMSK-modulated interfering signal.

For a BTS, the ISL is −93 dBm + C/Ic, where C/Ic is the cochannel interference ratio, given in Table 4.10 for the example of GSM-900 and GSM-850. The BLER performance is to be met for the three following cases:

- *Cochannel interferer*—the interfering signal is on the same RF channel as that desired (same ARFCN). The carrier to interference ratio is referred to as C/Ic.

- *First adjacent channel*—the interfering carrier is 200 kHz away from the desired signal RF channel. The carrier to interference is given by $C/Ia_1 = C/Ic - 18$ dB.

- *Second adjacent channel*—the interference signal is transmitted at 400 kHz from the desired carrier. The carrier to interference ratio is $C/Ia_2 = C/Ic - 50$ dB

Table 4.10
C/I Ratio for Cochannel Performance for Normal BTS

Type of Channel	GSM-900 and GSM-850 Normal BTS				
	Propagation Conditions				
	TU3 (No FH)	TU3 (Ideal FH)	TU50 (No FH)	TU50 (Ideal FH)	RA250 (No FH)
PDTCH/CS-1 (dB)	13	9	10	9	9
PDTCH/CS-2 (dB)	15	13	14	13	13
PDTCH/CS-3 (dB)	16	15	16	15	16
PDTCH/CS-4 (dB)	21	23	24	24	No requirement
USF/CS-1 (dB)	19	10	12	10	10
USF/CS-2 to 4 (dB)	18	9	10	9	8
PRACH/11 bits (dB)	8	8	8	8	10
PRACH/8 bits (dB)	8	8	8	8	9

If we take the example of CS-4 in GSM900, with the TU50/no FH propagation profile, we have the following conditions for a normal BTS:

- The desired ISL is −93 + C/Ic with C/Ic = 24 dB, so ISL = −69 dBm, for the cochannel, first adjacent, and second adjacent specifications.

- To fulfill the cochannel interference performance, the C/Ic is fixed to 24 dB, so the BLER performance is to be met for a signal level of −69 dBm, and a cochannel level of −93 dBm.

- In the first adjacent channel interference performance case, the wanted signal level is also equal to −69 dBm, and since C/Ia1 = C/Ic − 18 dB = 24 − 18 = 6 dB, the interferer level is −69 − 6 = −75 dBm.

- For the second adjacent channel interference specification, the wanted signal level is still equal to −69 dBm, with C/Ia2 = C/Ic − 50 dB = 24 − 50 = −26 dB, so the interferer level is equal to −69 − (−26) = −43 dBm.

For the MS, the C/I requirements are the same as for the BTS, but the ISL of the desired carrier for which the performance will be met is higher (by 0, 2, or 4 dB according to the type of MS).

Table 4.11
In-Band and Out-of-Band Definitions for the Blocking Characteristics

Frequency Band	MS	BTS
In band	915–980	860–925
Out-of-band (a)	0.1 to 915 MH	0.1 to 860 MHz
Out-of-band (b)	980 to 12,750 MHz	925 to 12,750 MHz

4.2.3.3   Blocking Characteristics

A blocking is an interfering signal at a high power level as compared with the desired signal, either in-band (which means situated in the receiver band), or out-of-band (out from the receiver band). Of course, the frequency bands corresponding to the in-band and out-of-band parts are dependent on the system (GSM-900, DCS-1800, and PCS-1900).

The blocking characteristics of the receiver refer to its capability of achieving a certain performance in the presence of a strong interferer in the in-band or one of the out-of-band frequency bands (see Table 4.11). This requirement exists in the case of the GSM circuit-switched service, and has also been defined for the GPRS system, from Release 99. No blocking characteristics were defined in Release 97/98 GPRS equipment.

If we take the example of GSM900, this requirement states that the sensitivity reference level performance (BLER requirements) will be met in the conditions here:

- A desired signal at frequency f_0, with a power level 3 dB higher than in the sensitivity requirement (see Section 4.2.3.1);

- A continuous sine wave blocking signal at a frequency f, which is an integer multiple of 200 kHz, and with a power level in dBm as specified in Table 4.12.

4.2.3.4   Intermodulation Characteristics

This performance requirement is valid for the GSM voice services, and has also been introduced in the GPRS requirements from Release 99. Release 97/98 transceivers are not requested to fulfill this requirement. The require-

Table 4.12
In-Band and Out-of-Band Blocking Signal Level (GSM-900)

Frequency Band	EGSM-900 Level of the Blocking Signal in dBm		
	Small MS	Other MS	BTS
In-band			
600 kHz = $\mid f - f_0 \mid$ < 800 kHz	−43	−38	−26
800 kHz = $\mid f - f_0 \mid$ < 1.6 MHz	−43	−33	−16
1.6 MHz = $\mid f - f_0 \mid$ < 3 MHz	−33	−23	−16
3 MHz = $\mid f - f_0 \mid$	−23	−23	−13
Out-of-band			
(a)	0	0	8
(b)	0	0	8

ment states that the sensitivity performance will be reached in the following conditions:

- A desired signal 3 dB above the sensitivity level requirement, at the frequency f_0;
- A continuous sine wave at the frequency f_1 added to this signal, with a level between −43 and −49 dBm according to the frequency band (see Table 4.13);
- A modulated GMSK signal on the frequency f_2, with the same level as the sine wave.

The frequencies are chosen such that $f_0 = 2 \cdot f_1 - f_2$ and $\mid f_2 - f_1 \mid = 800$ kHz.

Due to the third-order nonlinearity of the receiver, the intermodulation product of frequencies f_1 and f_2 generates a signal at frequency f_0. This results in a modulated interference signal that is added to the signal of interest. The effect is therefore similar to a cochannel interference on the received baseband signal. This point is further developed in Section 4.3.5.2.

Table 4.13

Power Levels of Signals on Frequencies f_1 and f_2 for Intermodulation Requirement

GSM-400, GSM-850, and GSM-900 small MSs	−49 dBm
DCS-1800 other than class 3 MSs, and PCS-1900 MS	
DCS-1800, PCS-1900 BTS	
DCS-1800 class 3 MS	−45 dBm
All other cases	−43 dBm

4.2.3.5 AM Suppression

The AM suppression characteristic is a receiver requirement [1] that concerns the GSM circuit-switched voice and data services, but also the GPRS services from Release 99 (Releases 97/98 GPRS equipment is not needed to fulfill this requirement).

In the AM suppression specification, the recommendations request fulfillment of the sensitivity performance with the following signals at the input of the reception chain:

- The desired signal, which is GMSK modulated with a power 3dB higher than the sensitivity level.

- A GMSK-modulated carrier situated in the receive band, at least 6 MHz away from the desired signal, with a power level as given in Table 4.14. Bursts are transmitted on this carrier, with a time delay between 61 and 86 bit periods relative to the bursts of the desired signal.

Table 4.14

AM Suppression Requirement

	MS (dBm)	BTS (dBm)
GSM-400	−31	−31
GSM-900	−31	−31
GSM-850	−31	−31
DCS-1800	−29 class 3 MS	−35
	−31 class 1 and 2 MS	
PCS-1900	−31	−35

The reason for this specification and the problems that arise in the receiver due to this requirement are discussed in a case study (see Section 4.3.4.2).

4.3   Case Studies

This part of Chapter 4 is dedicated to a presentation of the practical problems that arise in the implementation phase of the system. It provides an opportunity to explore in further detail the topics covered earlier. The first three cases relate to the PLL (channel coding and decoding issues, measurements), while the four subsequent ones deal with the RF physical layer (design of the transceiver, synthesizer).

4.3.1   Convolutional Coder

As seen in the previous sections, the channel coding schemes CS-1, CS-2, and CS-3 are based on a convolutional code, a frequently used mechanism to improve the performance in a digital radio communication system. This code is used to introduce some redundancy in the data bit sequence, to allow the correction (at the receiver end) of errors that may occur on these data bits during transmission, due to propagation conditions, interference, RF impairments, and so on.

A convolutional code is a type of error correction code in which each k-bit information symbol to be encoded is transformed into an n-bit symbol. In many cases, k is equal to 1, which means that each bit of the sequence to be encoded is input one by one in the coder, and each input bit produces a word (also called a symbol) of n bits.

In such a code, the n output bits are not only computed with the k bits at the input, as in a block code, but also with the $L-1$ previous input bits. The transformation is a function of the last L information symbols, where L is the constraint length of the code. This is the difference with block codes, for which only the input data bits are taken into account to compute the output bit sequence.

The ratio k/n is called the code rate. The degree of protection achieved by a code is increased when the code rate is decreased, which means that higher redundancy is introduced. A code is often characterized by as (n,k,L), where L is the constraint length of the code. This constraint length repre-

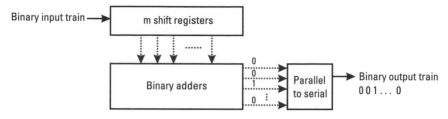

Figure 4.13 Circuit for a convolutional code.

sents the number of bits in the encoder memory that affect the generation of output bits. A convolutional code can be implemented as a k-input, n-output linear sequential circuit with input memory m, as shown in Figure 4.13.

The coder structure may be derived in a very simple way from its parameters. It comprises m boxes representing the m memory registers, and n modulo-2 adders for the n output bits. The memory registers are connected to the adders.

The rule defining which bits are added to produce the output bits is given by the so-called generator polynomials. For example, for a ratio 1/2 code, 2 polynomials are used, as shown for the code defined by G0 and G1 on Figure 4.14, with

$$G0 = 1 + D^3 + D^4$$

$$G1 = 1 + D + D^3 + D^4$$

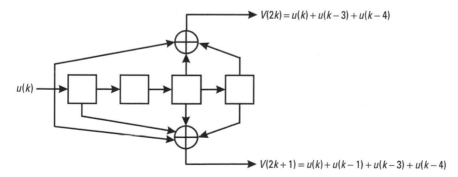

Figure 4.14 Circuit for the rate 1/2 code defined by G0 and G1.

In this notation, D represents the delayed input bit. If we denote by $u(n)$ and $v(n)$ respectively the input and output sequences of the coder, we have

$$v(2k) = u(k) + u(k-3) + u(k-4) \tag{4.6}$$

$$v(2k+1) = u(k) + u(k-1) + u(k-3) + u(k-4) \tag{4.7}$$

Usually, we have $u(k) = 0$ for $k < 0$. This code defined by G0 and G1 generates two output bits for every 1 input bit. Its constraint length L is equal to $m + 1$, which is 5 in this case.

For this example, suppose that the initial state of the registers is all zeros (0000), and that the sequence 1001 is presented at the input. The following operations are performed (see Figure 4.15):

- The first input bit "1" generates the two bits "11" on the output ($1 + 0 + 0 = 1$ and $1 + 0 + 0 + 0 = 1$), and the state of the registers becomes 1000, since the input bit one moves forward in the registers.
- The input is now the bit "0," which generates the pair of bits "01," and the new state is 0100.
- The next input is "0:" it generates the pair "00," with a shift in the registers to the state 0010.

A "1" is entered in the coder, so the output becomes "00."

In this example, the complete output sequence is therefore 11010000.

At the end of the coding process, for each coded block, a fixed sequence of bits is usually placed at the input of the coder, in order to bring the coder to a known state. This will be described further in the next section, since it is useful to the decoder. This fixed sequence is called the tail bit.

The codes used in GPRS are defined by the polynomials G0 and G1, presented above, that are also used for several other GSM logical channels (including TCH/FS and SACCH). It is interesting to notice that the coding schemes CS-1 to CS-3 are all based on the same generator polynomials, although not resulting in the same code. This is due to the fact that the puncturing technique is used to change the code rates. With the coder structure presented above, the achievable code rates are 1/2, 1/3, 1/4, 2/3, 2/5,

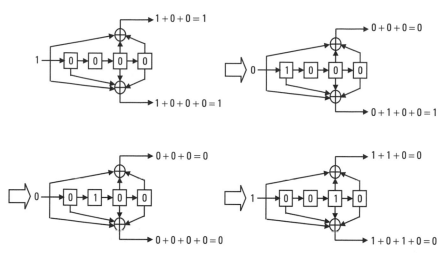

Figure 4.15 Example with the input bit sequence 1 0 0 1.

and so on, since *k* and *n* are integer values. These codes are often denoted as "mother codes." Puncturing is a method that allows the code rates to be changed in a very flexible manner. This is done in the transmitter by suppressing some of the bits of the coder output sequence, in particular positions. These positions are known to the receiver, which can replace the erased bits with zeros. The decoding process is then just the same as for the basic mother code. Many higher rated codes can therefore be formed, which allows for the provision of different degrees of protection, and adaptation of the bit rate to the available bandwidth.

Taking into consideration the CS-3 case, 338 bits are input to the 1/2 coder, resulting in 676 coded bits, and 220 of these bits are punctured, so that 456 bits are actually transmitted in a radio block. The overall rate of the code is in this case $338/(676 - 220) \approx 0.741$. The 220 punctured bits are such that among the 676 coded bits, the bit positions $3 + 6j$ and $5 + 6j$, for $j = 2, 3, ..., 111$ are not transmitted. Table 4.15 shows the code rates of the CS-1 to CS-3 codes.

4.3.2 Viterbi Decoding

There are several different approaches with respect to the decoding of convolutional codes. The Viterbi algorithm is most often used. This algorithm, introduced by Viterbi [2] in 1967, is based on the maximum likelihood prin-

Table 4.15
CS-1 to CS-3 Code Rates

Coding Scheme	Input Bits of the Coder	Output Bits	Punctured Bits	Overall Coding Rate
CS-1	228	456	0	228/456 = 1/2
CS-2	294	588	132	294/(588 − 132) = 0.645 ≈ 2/3
CS-3	338	676	220	338/(676 − 220) = 0.741 ≈ 3/4

ciple. This principle relies on intuitive behavior in comparing the received sequence to all the binary sequences that may have been coded, and choosing the closest one. Once the coded sequence closest to the receive sequence is identified, the input bits of the code can be deduced. Of course, the complexity of such an exhaustive comparison is not reasonable if we consider the implementation point of view, and it is the purpose of the Viterbi algorithm to reduce the number of comparisons, as will be explained later in this section.

First of all, it is convenient to introduce the maximum-likelihood concept. Let us assume that the sequence of information bits $u = (u_0, \ldots, u_{N-1})$ is encoded into a code word $v = (v_0, \ldots, v_{M-1})$, and that the sequence $r = (r_0, \ldots, r_{M-1})$ is received, on a *discrete memoryless channel* (DMC). This very simple channel is shown in Figure 4.16: The transmitted bits can either be transmitted correctly, with the probability $1 - p$, or with an error, with the probability p. This channel has no memory in the sense that consecutive received bits are independent.

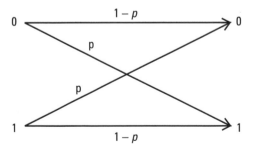

Figure 4.16 Discrete memoryless channel.

The decoder can evaluate the maximum-likelihood sequence as the codeword v maximizing the probability $P(r|v)$. On a memoryless channel, this probability is

$$P(r|v) = \prod_{i=0}^{M-1} P(r_i | v_i)$$

where $P(r_i|v_i)$ is the channel transition probabilityof the bit i, since the received symbols are independent from each other.

We can therefore derive that

$$\mathrm{Log}\left[P(r|v)\right] = \sum_{i=0}^{M-1} \mathrm{Log}\, P(r_i | v_i)$$

Indeed, if we suppose that $P(r_i|v_i) = p$ if $r_i \neq v_i$, and $P(r_i|v_i) = 1-p$ if $r_i = v_i$, then we have

$$\mathrm{Log}\, P(r|v) - d(r,v)\square \mathrm{Log}\, p + \left(M - d(r,v)\right)\square \mathrm{Log}(1-p)$$
$$= d(r,v)\square \mathrm{Log}\left(\frac{p}{1-p}\right) + M\,\square \mathrm{Log}(1-p) \tag{4.8}$$

where $d(r,v)$ is the Hamming distance between the two words r and v; that is, the number of positions in which r and v differ. For instance, the distance $d(100101,110001)$ is 2.

What is important to understand from (4.8) is that, on a DMC, maximizing the likelihood probability $P(r|v)$ is equivalent to minimizing the Hamming distance between the received word r and the possible transmitted coded sequence v. Indeed, the maximal value for a probability is 1, so $\mathrm{Log}\, P(r|v)$ maximal value is 0. THis characteristic is used for the decoder, as described in this section.

The decoding relies on a representation of the code called a *trellis*. A trellis is a structure that contains nodes, or states, and transitions between these states. The states represent the different combinations of bits in the registers. There are 2^{L-1} such combinations, L being the constraint length of the code.

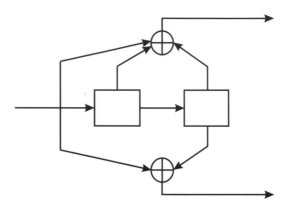

Figure 4.17 Example of 1/2 convolutional code.

For instance, for the code presented in Figure 4.17, based on the polynomials $g_1(D) = 1 + D + D^2$ and $g_2(D) = 1 + D^2$, there are four states: 00, 01, 10, and 11. The four possible states of the encoder are represented as four rows of dots. There is one column of four dots for the initial state of the encoder, and it is repeated for each step of the algorithm.

The coder in a given state changes to another state according to the input bit. A dashed arrow represents a 0 input bit, and a full line represents a 1 at the input of the encoder. For instance, if the coder is in the state 01, it will either change to the state 00, if the input bit is equal to 0, or 10, if the input bit is equal to 1. This is represented by the transitions, or arrows, in the trellis diagram.

For a 5-bit message at the input of the coder, for instance, there will be five steps in the algorithm. The labels on the transitions are the outputs of the coder. With the 1/2 code of our simple example, the transition between the states 01 and 00 is labeled 11. Indeed, if the state is 01, and the bit 0 is input to the coder, the output bits are $0 + 0 + 1 = 1$, and $0 + 1 = 1$ (remember that all the operations are performed modulo 2).

A trellis for this code is also shown in Figure 4.18. Note that once the transitions between the pair of states are made, this basic set of transitions can be repeated N times, N corresponding to the size of the input bits block. As we will see, each of these N repetitions of the trellis basic structure will correspond to a step in the algorithm.

A transition in the trellis is also denoted as a branch, and a series of several branches, between one state and another, is called a path in the trellis.

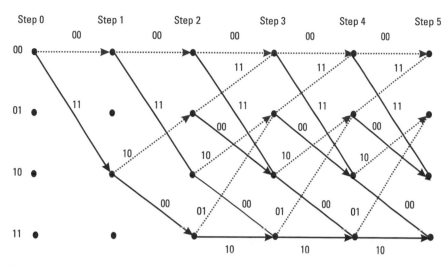

Figure 4.18 Associated trellis diagram for the example code.

Also notice that the initial state of the encoder is 00, so the arrows start out at this state. No transitions from the other states are represented at the beginning of the trellis. After two steps, all the states are visited, since all the combinations of bits are possible in the registers. The structure of the trellis diagram helps to compute the maximum-likelihood sequence. The principle of the Viterbi algorithm is to select the best path according to the metric given by the Hamming distance.

For the sake of example, let us suppose that the coded sequence 00 11 10 00 10 is transmitted, and that the receiver receives the sequence 00 11 10 01 10. Let us run the algorithm on this sequence to show how it can determine what the correct sequence is.

For each received pair of channel symbols, a metric is computed, to measure the Hamming distance between the received sequence and all of the possible channel symbol pairs. From step 0 to step 1, there are only two possible channel symbol pairs that could have been received: 00 and 11. The Hamming distance between the received pair and these two symbol pairs is called a branch metric.

The two first received bits are 00, so between step 0 and step 1, the branch metrics are calculated as follows:

- The transition between states 00 and 00 is marked with the label 00, so the branch metric is $d(00,00) = 0$.

The transition between states 00 and 10 is labeled 11, so the branch metric is $d(00,11) = 2$.

This process is then repeated for the transition between steps 1 and 2. The received sequence is then 11, and there are four transitions to explore:

- Between states 00 and 00 the branch metric is $d(11,00) = 2$.

- Between states 00 and 10 the branch metric is $d(11,11) = 0$.

- Between states 10 and 01 the branch metric is $d(11,10) = 1$.

- Between states 10 and 11 the branch metric is $d(11,00) = 2$.

The algorithm really starts after step 2, since there are now two paths that converge, for each state of the trellis.

We have seen that for each pair of received bits, the branch metrics associated with the transitions between states are calculated. In the algorithm, a path metric is also computed, as the sum of the branch metrics that compose the path. These cumulated metrics are calculated for each path that converges to a given node. Once these cumulated metrics are computed, the algorithm selects for each node the so-called survivor. A survivor is the path whose cumulated metric is the smallest, between the paths that merge to a given state.

In our example, the survivor, at step 3, for the state 01 is the path that comes from the state 10. The path that comes from the state 11 is removed. Table 4.16 gives, for each step, the estimation of the survivor path metric at each node of the trellis.

The idea is that each path in the trellis from state 00, at step 0, to one of the four states 00, 01, 10, and 11, at step n, represents a different sequence that may have been received. Indeed, there are 2^n paths of length n between state 0 and the four states at step n. The sequences represented to one path is the series of labels on each of its transitions.

What is important to understand is that for each state, one of the two paths that converge to this state is better than the other, with respect to the Hamming distance. When the minimum value between the two path metrics is chosen, this means that the path corresponding to this metric is selected, the other being removed. Instead of estimating the Hamming distance between all the possible sequences of n bits and the received sequence, the algorithm deletes half of the possible combinations at each step. It is indeed shown that for a given node, if the path P1 has a higher cumulated metric than the survivor P2, it will never be the maximum likelihood

Table 4.16
Survivor Path Metrics

	Received Sequence	State 00	State 01	State 10	State 11
Step 0	—	0	—	—	—
Step 1	00	$d(00,00) + 0 = 0$	—	$d(00,11) + 0 = 2$	—
Step 2	11	$d(11,00) + 0 = 2$	$d(11,10) + 2 = 3$	$d(11,11) + 0 = 0$	$d(11,00) + 2 = 4$
Step 3	10	$\text{Min}(d(10,00) + 2,$ $d(10,11) + 3) = 3$	$\text{Min}(d(10,10) + 0,$ $d(10,01) + 4) = 0$	$\text{Min}(d(10,11) + 2,$ $d(10,00) + 3) = 3$	$\text{Min}(d(10,00) + 0,$ $d(10,10) + 4) = 1$
Step 4	01	$\text{Min}(d(01,00) + 3,$ $d(01,11) + 0) = 1$	$\text{Min}(d(01,10) + 3,$ $d(01,01) + 1) = 1$	$\text{Min}(d(01,11) + 3,$ $d(01,00) + 0) = 1$	$\text{Min}(d(01,00) + 3,$ $d(01,10) + 1) = 3$
Step 5	10	$\text{Min}(d(10,00) + 1,$ $d(10,11) + 1) = 2$	$\text{Min}(d(10,10) + 1,$ $d(10,01) + 3) = 1$	$\text{Min}(d(10,11) + 1,$ $d(10,00) + 1) = 2$	$\text{Min}(d(10,00) + 1,$ $d(10,10) + 3) = 2$

sequence. This is clear since for the next steps, P2 will always be better than P1 with regard to the path metric.

In our example, at step 5, the survivors corresponding to each step are depicted in Figure 4.19.

This process of choosing the survivor for all the states is repeated until the end of a received sequence. Once this is reached, the trace-back procedure is performed. This procedure corresponds to:

- *The selection of the best survivor.* This is either the path that has the smallest metric among all four survivors, at the last step, or the survivor that corresponds to the known ending state. Indeed, if tail bits are used at the encoding, to bring the coder to a known state, the decoder does know which survivor to select. For instance, in our example, two bits equal to 0 will bring the coder to the state 00. If the receiving end knows that the state 00 is reached at the end of the sequence, it is able to perform the decoding more effectively (it will not chose the survivor from another state and thus will avoid errors in the end of the sequence).

- *The determination of the maximum likelihood transmitted sequence.* This is done by examining each branch of the trellis, to see if it corresponds to a 0 input (dashed line), or a 1 input (full line). The series of 0 and 1 represented by the arrows of the path give the decoded sequence.

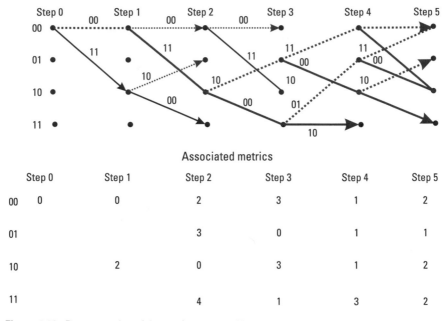

Figure 4.19 Representation of the survivors at step 5.

In our example, if we suppose that step 5 is the last step, we see that the best survivor is the one from state 01, since its Hamming distance is 1. This means that the received bit sequence differs from one bit to the sequence corresponding to this path (i.e., the sequence of the labels on each branch of this path). In this case, the transmitted sequence corresponding to the received bits 00 11 10 01 10 would be 01010. We also see that this input sequence produces the bits 00 11 10 00 10 in the encoder, which means that an error has occurred at the position 8 in the sequence, but the sequence was correctly decoded anyhow.

Of course, this is just an example, and the block size is usually longer than that, before the trace-back is performed (in the CS-3 case, for instance, 338 steps of the algorithm are executed, on a 16-state trellis). Note that the efficiency of the decoding process is increased if the number of steps is high.

The decoding method that we have seen in this example can be used in a similar way for any k/n code, and can be summarized as follows:

- Notations
    - N is the length of the data information sequence;
    - L is the constraint length of the code.

- For each step from $t = 0$ to $N - 1$
  - For each state from $s = 0$ to $s = 2^{L-1}$

    Estimation of the 2^k branch metrics that converge to the state s (Hamming distance between the received sequence and the sequence corresponding to the transition);

    Estimation of the path metrics for all the paths that converge to s;

    Determination of the survivor for state s: it is the path that has the minimum path metric, among all the paths that converge to s.
- Selection of either the best survivor among the 2^{L-1} states at stage $N - 1$ (the one that has the minimum path metric), or the survivor corresponding to the known ending state (if tail bits are used).
- Trace back to determine the input bit sequence corresponding to this best survivor.

The major interest of this algorithm is that its complexity is proportional to the number of states (number of operations proportional to $N \times 2^{L-1}$). This is a great complexity improvement compared with the 2^N comparisons that would be needed to explore all the possible received sequences.

The algorithm was described in this section for the DMC, that is, by supposing that the decoder input is a binary sequence, via hard decisions on the received samples. Convolutional codes are often based on soft decoding, which means that the input of the decoder is a sequence of "analog" samples, or soft decisions.

Examples of soft decisions would be:

- Binary samples ± 1 on an additive white Gaussian channel. The received samples r_k are Gaussian and uncorrelated. In this case, it can be shown that the metric to be used to compute the maximum likelihood of a path in the trellis diagram is the Euclidean distance. If a sequence r_k is received, the Euclidean distance between this sequence and a sequence of N binary symbols a_k in $\{\pm 1\}^N$ would be

$$\sqrt{\sum_{k=1}^{N}(r_k - a_k)^2}$$

- *Equalizer soft decisions.* In radio communication systems, the outputs of the equalizer are usually soft decisions on the received bits.

This means that instead of delivering ±1 values, the equalizer output is a sequence $y_k = b_k p_k$, where b_k is a binary symbol $\in \{\pm 1\}$, and p_k is proportional to the probability associated with this symbol. The idea is that the equalizer provides for each received bit, an estimation of the probability that this bit is the original bit that has been sent. A soft decision is therefore a reliability value, which may be used to increase the decoding capability. In such a case, the metric used in the Viterbi algorithm can be.

$$\sum_{k=1}^{N} (y_k a_k)$$

If y_k and a_k have the same sign, the metric is increased; otherwise, it is decreased. In this case, the Viterbi algorithm is used to select the path that maximizes this metric.

4.3.3 MS Measurements

This section describes the estimation principles for the measurements that are reported by the MS to the network.

RXLEV

As seen in the previous sections, this parameter corresponds to a received signal strength measurement on:

- The traffic channels if the MS is involved in a circuit-switched communication (GSM speech or data), or in GPRS packet transfer mode;
- On the BCCH carriers of the surrounding cells, for circuit- and packet-switched modes;
- On carriers indicated for interference measurements, in packet transfer or packet idle modes (these measurements are mapped on the RXLEV scale).

This received signal strength is evaluated on the baseband received I and Q signals, usually after the analog-to-digital conversion. The estimation of the power in dBm is done on samples, (I_k) and (Q_k), quantified on a certain number of bits:

$$10 \cdot \log \left[\frac{1}{N} \sum_{k=1}^{N} I_k^2 + Q_k^2 \right] + 30 - G \tag{4.9}$$

where N is the number of samples used for the estimation, and G is the gain of the RF receiver in decibels, including the analog-to-digital converter gain. To get the result in dBm, the power estimation is converted to decibels ($P_{dB} = 10 \cdot \log(P_{Watt})$), and the term 30 is added ($P_{dBm} = 10 \cdot \log(P_{mWatt}) = P_{dB} + 30$). Note that the receiver gain must be deducted from the result to estimate the received power level at the antenna.

Several estimations in dBm, taken from different time slots, are averaged for the computation of the RXLEV. Usually, each estimation is estimated on one received burst, or less than one burst in the case of the surrounding cells monitoring. Indeed, it may happen for some multislot cases that the MS must reduce its measurement window, in order to increase the time left for channel frequency switching. The precision will of course depend on the number of samples N used for each estimation, and on the number of estimations that are averaged.

The measured signal level is then mapped onto a scale between 0 and 63, and reported to the network.

RXQUAL

As seen in Section 3.3.1.5, this parameter is an estimation of the BER on the downlink blocks at the demodulator output (no channel coding). This estimation is obtained by averaging the BER on the blocks intended for the MS that are successfully decoded, in the sense that no error is detected by the CRC check. On these correctly decoded blocks, it is assumed that the convolutional code, for CS-1 to CS-3, is able to correct all the errors that have occurred on the physical link. This is a reasonable assumption since the probability that an error on the decoded bit remains undetected by the CRC decoder is very low.

Since we assume that the errors are all corrected by the convolutional code, a simple way of estimating the BER on the demodulated bits of one block is to proceed as follows:

- Once a block, mapped on four bursts on a given PDCH, is received, the soft bit decisions delivered by the equalizer are used for the convolutional decoding.

- The CRC of the decoded block is checked, and if no error is detected, the block is used in the BER estimation process, if it is addressed to the mobile; otherwise, it is skipped.

- If the decoded block is kept, it is reencoded by the convolutional encoder, to produce the coded bits, just as in the transmitter.

- These coded bits are compared with the hard decisions that are delivered by the equalizer. The number of detected errors is simply divided by the length of the coded block—that is, 456 bits—to produce an estimate of the BER.

The estimated BER on several blocks is then averaged to compute the RXQUAL, which is a value between 0 and 7. This method is summarized in Figure 4.20. Note that this estimation is not possible on the blocks that are not correctly decoded (CRC check), since it cannot be assumed on these blocks that the convolutional decoder has corrected all the errors (otherwise the CRC check would be okay), and therefore the accuracy cannot be met.

This method cannot be used if the code is CS-4, since CS-4 has no convolutional coding. For this reason, reporting of a precise estimate of the BER is not requested in CS-4, and the MS is allowed to report RXQUAL = 7 (this corresponds to a BER greater than 12.8%).

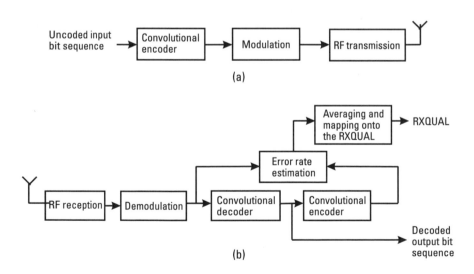

Figure 4.20 BER estimation for the RXQUAL measurement: operations in the (a) transmitter and (b) receiver.

4.3.4 RF Receiver Structures

This section is a summary of the architectures commonly in use in the MS and BTS receivers. Of course, the implementation of the receiver is not described in the standard, but it will be designed by the manufacturers to meet the specified performance. This case study presents the basics of RF receiver architectures, in order to show the different constraints of the recommendations and their impact on the practical implementation. It is suggested that the reader first review Section 4.2 covering the GPRS RF requirements. For a detailed presentation of the RF architecture and implementation issues, refer to [3].

4.3.4.1 Intermediate Frequency Receiver

The purpose of the RF receiver is obviously to receive the RF modulated signal, and to transpose this signal to the baseband I and Q signals. These signals are then digitized (in the ADCs), and treated by a set of demodulation algorithms (CIR estimation, time and frequency synchronization, equalizer, decoding, and so on).

The *intermediate frequency* (IF) structure, or superheterodyne receiver, is shown in Figure 4.21. The receiver front-end contains a band selection filter (to select all the downlink bands of the system), and a *low noise amplifier* (LNA). Two stages are then used for the down-conversion of the RF signal to the baseband I and Q signals. The basic principle is the conversion of the RF carrier to a fixed IF by mixing it with a tunable LO. A bandpass filtering is then applied at this IF frequency, prior to a second conversion into baseband.

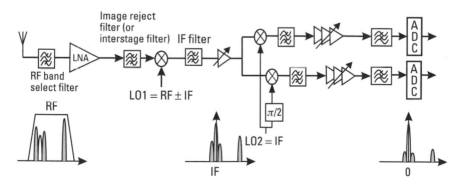

Figure 4.21 IF receiver architecture.

In the first stage, the mixing of the RF input signal to the RF ± IF LO has the effect of transposing the RF signal down to the IF. This is due to the fact that the LO signal, $A_0 \cdot \cos(2\pi f_{LO}t)$, and the RF wanted signal $A(t) \cdot \cos[2\pi f_{RF}t + \varphi(t)]$ are placed at the mixer's input. f_{LO} and f_{RF} are the LO and RF frequencies, A_0 is the amplitude of the LO sinusoidal signal, $A(t)$ is the *amplitude modulation* (AM) (with the ideal GMSK signal, it is constant), and $\varphi(t)$ is the phase modulation signal.

The result of this product is

$$A_0 \cdot A(t) \cdot \cos 2\pi f_{LO}t \cdot \cos[2\pi f_{RF}t + \varphi(t)]$$
$$= \frac{1}{2} \cdot A_0 \cdot A(t) \cdot \cos[2\pi(f_{RF} - f_{LO})t + \varphi(t)] \qquad (4.10)$$
$$+ \frac{1}{2} \cdot A_0 \cdot A(t) \cdot \cos[2\pi(f_{RF} + f_{LO})t + \varphi(t)]$$

If, for instance, $f_{LO} = f_{RF} - f_{IF}$ the result is

$$\frac{1}{2} \cdot A_0 \cdot A(t) \cdot \cos[2\pi f_{IF}t + \varphi(t)]$$
$$+ \frac{1}{2} \cdot A_0 \cdot A(t) \cdot \cos[2\pi(2f_{RF} - f_{IF})t + \varphi(t)]$$

The band pass filtering of this signal removes the second term, and the result is a modulated signal transposed to an IF central frequency. By changing f_{LO}, all the RF channels can be converted to a constant IF. The second stage of this architecture is needed to make the downconversion from IF to baseband (around 0 Hz central frequency). In order to do so, each of the IF LO in-phase $A_1 \cdot \cos(2\pi f_{IF}t)$ and quadrature

$$A_1 \cdot \cos\left(2\pi f_{IF} + \frac{\pi}{2}\right) = -A_1 \cdot \sin(2\pi f_{IF}t)$$

signals are mixed with the IF modulated signal.

This produces the two signals $I(t)\alpha A(t) \cdot \cos\varphi(t)$ and $Q(t)\alpha A(t) \cdot \cos\varphi(t)$. Note that with $f_{LO} = f_{RF} + f_{IF}$, the result is similar.

$I(t)$ is called the in-phase signal, and $Q(t)$ the quadrature signal. The complex signal $I(t) + j \cdot Q(t)$ is the baseband received signal, also called the

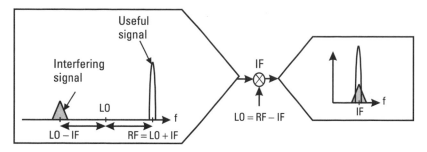

Figure 4.22 Problem of image frequency in IF architecture.

complex envelope of the RF signal. Note that the RF signal can be written $I(t) \cdot \cos(2\pi f_{RF}t) - Q(t) \cdot \sin(2\pi f_{RF}t)$.

One important issue with the IF receiver architecture is the problem of the image frequency. This problem is illustrated in Figure 4.22. In the first stage of the receiver, the RF signal is converted to IF, since, in our example, $f_{LO} = f_{RF} - f_{IF}$. In this case, the signal that is present around $f_{LO} - f_{IF}$, at the input of the receiver, is also downconverted to IF and added to the modulated signal. This signal may for instance be another carrier of the system, or an interference out of the reception band, depending on the choice of IF frequency. After downconversion to baseband, this unwanted signal is added to the modulated signal, resulting in a cochannel interference, thus degrading the performance of the receiver.

For this reason, the IF LO frequency is often chosen to be large enough so that the image frequency ($f_{LO} - f_{IF}$ in our example) is filtered out by the RF reception band filter, placed at the receiver front end. Such filter allows the signal placed at the image frequency to be partially removed, resulting in an improved signal to interference ratio at baseband.

As shown in Figure 4.21, an image reject filter, also called interstage filter, may also be placed before the RF mixer stage, in order to filter out the signal present at the image frequency. Note that between the RF and IF mixing stages, an IF filter is usually used to perform the channel filtering. Such a filter is usually very selective, and allows the suppression of adjacent channel interference and wideband noise.

In the IF architecture, the current technology does not permit integration of the interstage and IF filters, which is the major drawback of this scheme. Indeed, it requires costly and bulky external filters.

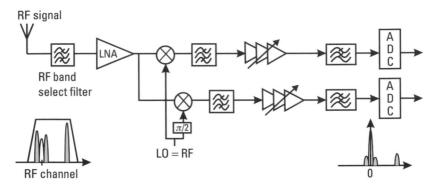

Figure 4.23 Zero-IF receiver architecture.

4.3.4.2 Zero-IF Receiver

The second classic RF receiver scheme is the zero-IF structure, or direct conversion receiver, which is based on a single downconversion stage (see Figure 4.23). This means that there is only one LO, with $f_{LO} = f_{RF}$, so that the I and Q signals are delivered at the output of the RF mixers. Note that the receiver front-end is similar to the IF case.

The benefit of having a single downconversion stage is of course a reduction in the components number, as compared with the IF scheme. Indeed, the interstage and IF filters are not required in this structure. For this reason, this architecture is one of the most cost-effective solutions, well suited for a mobile phone design, for which the cost and surface are critical issues. The drawback of this architecture is the so-called dc offset generated together with the baseband-modulated signal. The dc offset is a constant signal that is added to the signal of interest. This offset may be caused by several sources, as depicted in Figure 4.24.

One source is the LO leakage due to the limited isolation between the LO and the input of the mixer and LNA [Figure 4.24(a)], which is mixed with the LO signal itself. The leakage may also reflect on external obstacles before this self-mixing effect occurs [Figure 4.24(b)]. Similarly, there may be some self mixing of a high-power interference signal (blocking), as shown in Figure 4.24(c), leaking from the LNA or mixer input to the LO. This means that a small portion of the interference power is coupled onto the LO. The mixing of this leakage signal with the interfering signal itself results in a dc component.

All these phenomena have the same effect: two signals at the same RF frequency are mixed together and produce a constant, or dc, on the baseband

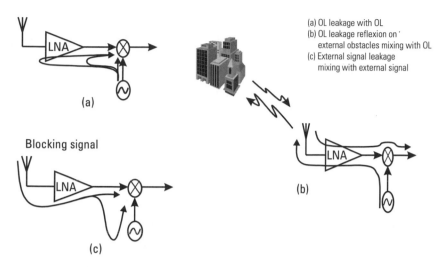

Figure 4.24 The dc offset sources in the ZIF receiver:

I and Q. In order to cope with the dc problem, analog solutions are possible, based on a dc compensation on the I and Q baseband signal. However, such solutions are not fully efficient in a context of varying radio conditions. Software solutions are therefore often required, to compensate for the residual dc on the digitized I and Q signals.

For Figure 4.24(c), one problem is that the blocking signal might not be synchronized with the receive burst. One can think of a carrier of the GSM system in the same band of operation, and therefore not filtered by the receive band filter. This carrier is not necessarily synchronized with the received channel frequency (different BTSs are not synchronized with one another): this raises the problem of a dc step occurring during the received burst (the beginning of the interfering carrier burst is not synchronized with the beginning of the desired carrier burst). The receiver has no information on the position and level of this step, and therefore complex digital processing techniques are needed to remove the step.

This problem arises in the conditions of the AM suppression characteristics (see Section 4.2.3.5). In this case, the RF mixers detect the envelope of the blocking signal and reproduce it on their output (see Figure 4.25), due to the RL-to-LO coupling and second-order nonlinearity distortion. Digital signal processing techniques may therefore be required to detect the step, to find out its position in the receive burst and its level. It is then possible to suppress this dc step. These problems concerning the dc offset may require the increase of the ADC's dynamic range, to avoid their saturation.

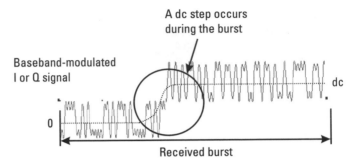

Figure 4.25 AM problem.

Another drawback of the zero-IF scheme compared to the IF solution is the possible degradation of the receiver *noise figure* (NF), which also means a degradation of the sensitivity level. This is due to the fact that the IF architecture performs better in terms of selectivity (filtering of the signal to remove the noise, interference, and blocking signals) in the IF stage. Therefore, the RF filtering prior to the LNA may be relaxed in the IF scheme compared to the zero-IF one. This distribution of the filtering allows the insertion loss of the receiver band filter to be reduced, leading to a better noise figure (see Section 4.3.5.1).

4.3.4.3 Near-Zero-IF Receiver

To avoid the problem of the dc offset, an interesting architecture is the low IF, or near-zero IF, which is actually a special case of the superheterodyne receiver (see Figure 4.26). In this scheme, the RF mixers' downconversion stage is used to translate the RF signal to IF, and the IF signal is directly digi-

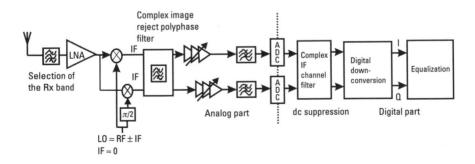

Figure 4.26 Near-zero-IF architecture.

tized and converted afterward to baseband in the digital part of the system. The IF will therefore be sufficiently low to allow a practicable ADC implementation (the ADC sampling frequency will be at least two times greater than the maximum input signal frequency, to respect the Shannon theorem, but the oversampling ratio is at least four in practical cases). The dc offset can be removed at this stage, prior to the second downconversion to baseband, which is done in the digital part of the system.

The image frequency rejection is a real issue in this method, since the image frequency is within the receive band, so it is not filtered out by the front-end filter (Rx band selection filter). Therefore, if the image is not rejected, the interference caused on the baseband signal may be very high, often higher than the desired signal itself. For instance, in an implementation where the IF is equal to 100 kHz, the image frequency is 200 kHz away from the RF carrier that needs to be demodulated, so it is an adjacent channel interference. This interference is specified 9 dB higher than the signal of interest in the GSM adjacent channel requirement. It is therefore very important to remove the image in the mixers, or with a complex polyphase filter after the first downconversion to IF. This filter acts as an all-pass filter for the positive frequencies, and as a bandstop filter for the negative frequencies. The efficiency of this filter may be dependent upon the quadrature accuracy (only low phase and amplitude mismatches are tolerated). Note that a great part of the channel filtering is usually achieved in the digital part of the system, which requires an important ADC dynamic range (interfering signals are not completely removed prior to the ADC stage). A detailed description of this scheme is given in [4].

4.3.4.4 Summary

The various receiver architectures are summarized in Table 4.17.

4.3.5 RF Receiver Constraints

4.3.5.1 NF

One of the most important specifications in a receiver is the sensitivity it is able to achieve, that is, the minimum level of signal power that can be received with an acceptable *signal-to-noise ratio* (SNR), ensuring a sufficient level of BER. The smaller the BTS and MS sensitivity levels, the better the performance of the whole system will be, since the link quality will be better in areas covered with large cells.

Table 4.17
RF Receiver Architectures

	Advantages	Drawbacks
IF	Good channel selectivity	Not the most effective in terms of cost and area; image problem
Zero IF	Cost-effective: reduction of the number of external components	The dc offset compensation and increase of the ADC dynamic range required
Near-zero IF	Same as zero IF, without the dc offset compensation problem	Digital complexity increases due to the downconversion from low IF to baseband; high ADC dynamic range; requires accurate quadrature

Factors influencing the sensitivity level are:

- The receiver NF, which is a measure of the SNR degradation at the output of the analog receiver, compared to the input;
- The channel filtering, which modifies the SNR, since it partly removes the noise;
- The performance of the baseband demodulation algorithms, which is characterized by BER curves (or BLER, see Section 4.2.3) versus SNR (the SNR is often under form of E_b/N_0, that is, the ratio of the signal energy during a bit period relative to the energy of the noise during this same period);
- RF imperfections;
- Imperfections in the digital signal processing part.

Knowing the capability of the demodulation algorithms to achieve the level of performance required in the recommendations (BER, BLER, and so on), in the presence of RF and digital signal processing impairments, the RF system designer can derive the NF needed in the receiver.

Below we analyze one by one the factors listed above.

Receiver NF

The NF is a measure of the SNR degradation as the signal passes through the system. It is determined by measuring the ratio of the SNR at the output to that at the input of the receiver.

$$NF = \frac{S_{\text{in}}/N_{\text{in}}}{S_{\text{out}}/N_{\text{out}}} \qquad (4.11)$$

where:

S_{out} and N_{out} are the wanted signal and noise powers at the output of the RF reception chain;

S_{in} and N_{in} are the wanted signal and noise powers at the input of the RF reception chain.

The overall receiver *NF* may be calculated from the *NF*s of the different elements of the system, with the Friss equation:

$$NF = NF_1 + \frac{NF_2 - 1}{G_1} + \frac{NF_3 - 1}{G_1.G_2} + \cdots + \frac{NF_N - 1}{G_1.G_2.\cdots.G_{N-1}} \qquad (4.12)$$

In this equation, NF_i and G_i represent the NF and gain of stage i. The NF is usually expressed in decibels: $NF_{\text{dB}} = 10 \cdot \log(NF)$.

What can be observed from (4.12) is that the higher the gain in the first stages, the lower the overall NF. Indeed, for the same total gain $G_1 + G_2 + \ldots + G_{N-1}$, the receiver NF will be lower if G_1 and G_2 are higher. A high gain in the receiver front end has the effect of "masking" the NF of the following stages. This is why, for instance, in the zero-IF architecture, the overall NF may not be as good as in the IF architecture: in the IF architecture, the insertion loss due to the front-end band selection filter may be decreased, because an important part of the filtering is performed at the at the IF.

The NF allows a convenient correspondence between the receiver ISL and the E_b/N_0 at the digital demodulator input. At the input of the receiver, the SNR is equal to the power of the RF signal of interest, relative to the thermal noise. A body at absolute temperature T (in degrees Kelvin) emits electromagnetic waves at a power level of kT W/Hz, where k is Boltzmann's constant. The thermal noise power is therefore $kT = -174$ dBm/Hz at room temperature ($k = 1.38\ 10^{-23}$ J $\cdot$ K^{-1}, and $T = 293$K). This thermal noise is usually integrated over a bandwidth B, corresponding to the equivalent noise bandwidth of the receiver system.

The energy per bit is $E_b = S_{\text{out}}/R$, where R is the symbol rate ($R = 270.83333\ 10^3$ baud), and $N_{\text{out}} = N_0 \cdot B$, where N_0 is the noise energy (in watts per hertz) at the output of the RF system.

Therefore, we have

$$\left(S_{out}/N_{out}\right)_{dB} = \left(E_b/N_0\right)_{dB} + 10\cdot\log\left(\frac{R}{B}\right)$$

Moreover, since $(N_{in})_{dB} = 10\cdot\log(kTB)$, we derive the following:

$$
\begin{aligned}
\left(NF\right)_{dB} &= S_{in} - 10\cdot\log\left(kTB\right) - \left(E_b/N_0\right)_{dB} - 10\cdot\log\left(\frac{R}{B}\right) \\
&= S_{in} + 174 - \left(E_b/N_0\right)_{dB} - 10\cdot\log\left(R\right) \\
&= S_{in} - \left(E_b/N_0\right)_{dB} + 119.6
\end{aligned}
\tag{4.13}
$$

These calculations are summarized in Figure 4.27.

Channel Filtering

Depending on the channel filter, the noise bandwidth B, and therefore the E_b/N_0, may be different: $(S_{out}/N_{out})_{dB} - 10\cdot\log(R/B) = (E_b/N_0)_{dB}$.

Demodulation Algorithm

To meet the BLER performance required in the 05.05 specification (see summary of RF receiver requirements in Section 4.2.3), a certain level of E_b/N_0 is needed at the digital demodulator input. This level is different according to the type of logical channel (PDTCH, PRACH), channel coding scheme (CS-1 to CS-4), and propagation conditions (static, TU50, RA250, HT100,

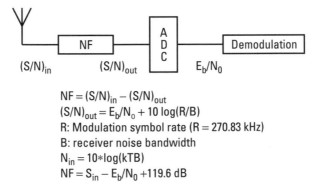

Figure 4.27 NF calculation.

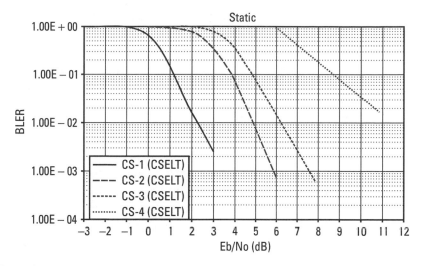

Figure 4.28 Demodulation performance BLER versus E_b/N_0, static channel. (*From:* [5].)

and so on). Also, according to the demodulation algorithms performance, there exist some variations on the E_b/N_0 required to reach the requested level of BLER.

As an example, Figure 4.28 shows the PDTCH CS-4 BLER versus E_b/N_0 of a receiver, for a static channel. This curve is taken from [5]. This kind of curve is usually derived by simulation means, with an ideal GMSK signal passed through a propagation channel simulator, with additive Gaussian noise (to set the wanted level of E_b/N_0) followed by the receiver filter and demodulation algorithms.

RF and Baseband Imperfections

Knowing the E_b/N_0 level required to reach the BLER performances, the ISL at the receiver for which the performance will be met (sensitivity level), the RF system designer can derive the maximal NF that can be tolerated in the receiver.

In order to do so, some margins will be taken in the assumed E_b/N_0. Indeed, according to the receiver implementations, several impairments can lead to a performance degradation. Among these impairments are the following:

- Quantification noise due to the ADCs;
- Clipping due to a saturation on the ADCs;

- I/Q imperfections such as phase mismatch (the quadrature in the receiver is not exactly $\pi/2$ or amplitude mismatch (the amplitude gain on I and Q may not be exactly the same);
- Group delay distortion in the receiver chain filters;
- Phase noise that is generated by the synthesizer;
- A dc offset on I and Q signals;
- Demodulation algorithms fixed-point implementation: the digital signal processor that performs the demodulation algorithms operates on finite length bit words. For instance, the multiplication-accumulations operations may be quantified on 16 bits.

Due to all these impairments, *implementation margins* (IM) are usually taken into account. The assumption that was used to derive the 05.05 sensitivity performance figures was a 2-dB IM.

The system designer may therefore take the following approach:

- Perform simulations to characterize the performance of the receiver in terms of required E_b/N_0, in an ideal environment (that is, without considering the impairments inherent to the receiver), for each logical channel and propagation condition, and for each coding scheme.
- Derive the margins on these performances to anticipate a degradation due to specific elements of the chosen RF and baseband implementations.

Equation (4.14) gives the NF constraint including an IM.

$$(NF)_{dB} = S_{in} - (E_b/N_0)_{dB} - (IM)_{dB} + 119.6 \qquad (4.14)$$

With this equation, and given the E_b/N_0 level of the demodulator to reach the BLER = 10%, for each case of the sensitivity level performance requirements (S_{in} is different depending on the coding scheme and channel profile), it is possible to derive an NF mask, giving for each ISL (between −102 and −88 dBm for a GSM-900 small MS, for instance) the maximal tolerated NF.

The NF constraint as described above is derived from the sensitivity level performance requirement. Nevertheless, it is to be understood that for the cochannel interference requirements, the ISL varies between −67 dBm and –82 dBm according to the coding scheme and propagation profile, for a GSM-900 small MS, for instance. In this requirement, a modulated interference signal is added to the signal of interest, and the 10% BLER performance is to be fulfilled for a given C/Ic ratio, which is also dependent upon the coding scheme and channel profile. The values for these C/Ic ratios in the 05.05 were derived assuming that a very good SNR was achieved. Thus, the results presented in the GSM 05.50 are based on an E_b/N_0 of 28 dB. This means that to fulfill the cochannel interference requirement, an E_b/N_0 of 28 dB may be needed in the range of –67 dBm to –82 dBm (in our example of small MS, GSM-900). Of course, if the demodulator reaches better performance it may be possible to still fulfill the requirement at a lower E_b/N_0.

4.3.5.2 Linearity

Linearity is an important issue in the receiver design. It defines the compression that is tolerated on the received signal. The nonlinear characteristic inherent in electronic devices such as amplifying and mixing circuits leads to undesired responses that are produced with the desired signal. Figure 4.29 represents the transfer function $F(x)$ of a nonlinear stage, where x is the input signal.

From this expansion, we see that higher-order distortion ($b \cdot x^2 + c \cdot x^3$ + ...) is added to the desired amplified input signal ($a \cdot x$). In a receiver design, the third-order distortion is a very important specification. This is due to the fact that when two signals at frequency f_1 and f_2 are input to a third-order nonlinear device, this produces intermodulation products at frequencies $n \cdot f_1 \pm m \cdot f_2$ with $n + m = 3$.

For instance, it generates an unwanted response at frequency $2 \cdot f_1 - f_2$. In the intermodulation characteristics specification (see Section 4.2.3.4), a modulated wanted signal at frequency f_0 is placed at the input of the receiver together with a continuous sine wave at the frequency f_1 and a modulated GMSK signal at the frequency f_2. The frequencies are chosen such that $f_0 = 2f_1 - f_2$ and $|f_2 - f_1| = 800$ kHz.

In such conditions, the unwanted response is added at the same frequency as the signal of interest, and it acts as a cochannel perturbation.

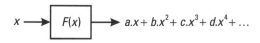

Figure 4.29 Nonlinear response.

In this section we show how the third-order nonlinearity can be quantified, and how it is related to the level of interference generated by the intermodulation product.

First of all, let us give a definition of the *third-order intercept point* (IP3), which quantifies the order three nonlinearity. Figure 4.30 shows the power response of the desired signal through the system, where P_{in} and P_{out} represent the input and output powers, in dBm. We see that the gain of the system expressed in decibels is G_{dB}, and that the output signal increases linearly with the input. With the notations of Figure 4.29, G_{dB} is equal to $10 \cdot \log(a)$. A second response is added, corresponding to the third-order term $c \cdot x^3$. In decibels, this is represented as $10 \cdot \log(c \cdot x^3) = 10 \cdot \log(c) + 3 \cdot 10 \cdot \log(x)$, and therefore it is a line of slope 3 if represented in logarithm scale. This second response crosses the y-axis for $P_{out} = 10 \cdot \log(c)$.

The two responses shown on this figure intercept at a theoretical point that is called the IP3 of the stage, expressed in dBm. It may be referenced from either the input ($IP3_{in}$) or output ($IP3_{out}$) of the stage: $(IP3_{out})_{dBm} = (IP3_{in})_{dBm} + G_{dB}$.

This theoretical fictive point allows us to quantify the third-order nonlinearity of the receiver. It is directly related to the coefficients a and c, which

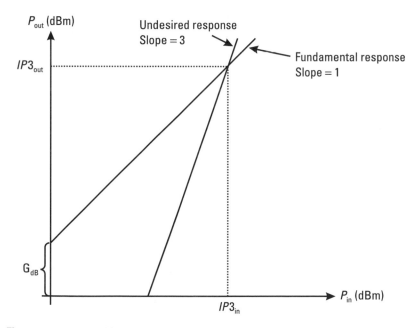

Figure 4.30 IP3 definition.

can also be used to characterize this third-order nonlinearity. The lower the IP3, the greater the nonlinearity distortion of the system.

The IP3 is usually used to specify a receiver design, since it allows convenient calculations. For instance, if each stage of a receiver chain is characterized by an IP3, the total input IP3 (*IP3in*) of the receive chain is calculated as follows:

$$\frac{1}{IP3in} = \frac{1}{IP3in_1} + \frac{G_1}{IP3in_2} + \frac{G_1 \cdot G_2}{IP3in_3} + \cdots + \frac{G_1 \cdot G_2 \cdots G_{N-1}}{IP3in_N} \qquad (4.15)$$

where $IP3in_i$ and G_i represent the input IP3 and gain of the ith stage of the receiver, in linear scale: $IP3in_{\mathrm{dBm}} = 10 \cdot \log(IP3in_i) + 30$, and $G_{i\mathrm{dB}} = 10 \cdot \log(G_i)$.

A simple observation with respect to (4.12) and (4.15) shows that there is a trade-off in the receiver design between the NF and the IP3 performance. Indeed, the higher the gain in the receiver front end, the lower the NF, but also the lower the IP3.

As stated before, in the conditions of the intermodulation characteristic requirement, a modulated interference signal is added to the signal of interest. The effect is therefore similar to a cochannel interference on the received baseband signal. The C/I ratio at baseband, noted

$$\left(\frac{C}{I}\right)_{BB}$$

is directly linked to the IP3*in* characteristic of the receiver as follows:

$$IP3in_{\mathrm{dBm}} = \frac{1}{2}\left(3 \cdot I_{RF} - ISL + \left(\frac{C}{I}\right)_{BB}\right) \qquad (4.16)$$

where *ISL* is the ISL expressed in dBm (3 dB above sensitivity performance in the requirement), and I_{RF} is the level of the RF interfering signal (in dBm).

Therefore, knowing the $(C/I)_{BB}$ performance of the demodulator [the $(C/I)_{BB}$ that leads to the 10% BLER, for instance], it is possible to derive the minimum IP3 constraint that should be respected in the receiver.

4.3.5.3 AGC Constraints

The AGC is an important function in the receiver; it adapts the gain of the different stages according to the receiver ISL. For GSM-900 speech services, for instance, the ISL for the MS can vary between –102 dBm and –15 dBm.

This 87 dB dynamic range is tremendous; it cannot be handled directly by the digital baseband section. Rather, the RF part must be able to adapt its total gain, so that the dynamic range in the digital baseband is reduced. In order to do this, the receiver usually makes an estimation of the received signal level at the digital baseband, and after an averaging on the last received bursts, it is able to estimate the RF signal level at the antenna and therefore to set the correct gain needed in the receiver (see Figure 4.31).

The averaged estimations of the received signal level are compared with a threshold, and the difference between the estimated level and the threshold, Δ, is used to predict the gain of the next receive slot. If the threshold Δ is positive, the gain is decreased; if it is negative, the gain is increased. With this process, the signal variations in the baseband are low, because the dynamic range of the ISL is handled by the gains of the overall receiver chain. The AGC strategy may be stored in the form of a table that contains for each Δ value the corresponding gain setting.

For a practical AGC implementation (that is the derivation of the gain settings for each ISL), many parameters must be chosen, such as the number of samples in the averaging of the received signal measurements, the averaging method (averaging window size, use of a forgetting factor, and so on), and the choice of the threshold. Also, it is important to note that the choice of the AGC threshold is strongly related to the ADC dynamic range.

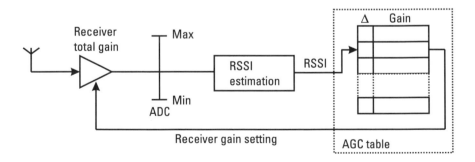

Figure 4.31 AGC loop mechanism.

These parameters are derived taking into account a number of constraints:

- The NF and IP3 constraints, described in Sections 4.3.5.1 and 4.3.5.2, which are dependent upon the receiver gain. For a given ISL, sufficient gain should be set in order to respect the NF constraint corresponding to that ISL, but not too much gain, to respect the IP3 constraint.

- The power control in downlink (see Section 4.1.3.2), which implies that the MS has only partial information of the transmitted downlink power. In particular, for a multislot mobile, it is usually not possible to switch the gain settings from one receive slot to the other in the same TDMA frame.

- The fading conditions, which produce variations on the received signal from one burst to the other, or during the burst.

- The dc offset that may be generated in the RF part of the system (for instance, in the zero-IF receiver structure presented in Section 4.3.4.2).

- The receiver channel filtering, on the adjacent channels and on the blocking signals.

- The interference measurement dynamic range. (See Section 4.1.4.2.)

- The minimum digital resolution needed for the demodulator to operate correctly. If the signal levels are not quantified on a sufficient number of bits, a significant degradation appears on the demodulation performance.

4.3.6 Transmitter Design

The purpose of the transmitter is to convert a baseband signal, at the output of the GMSK modulator, to an RF-modulated carrier. The architectures commonly in use for GSM and BTS mobiles (RF transmitter architectures are described in [3]) may also be used for the GPRS system. The new constraint for GPRS is the multislot power versus time mask (see definition in Section 4.2.2).

When the bursts in two or more consecutive time slots are transmitted in the same TDMA frame, the single-slot power versus time mask will be fulfilled during the useful part of each burst and at the beginning and the end of the series of consecutive bursts. In between two bursts, two cases are possible. The output power during the guard period will not be greater than the level in the useful part of time slot one, if this level is higher than the level during the useful part of time slot two +3 dB. If it is not the case, then the output power limit during the guard period is the level during the useful part of time slot two +3 dB.

As explained earlier in this chapter, the power control in the uplink is independent for each time slot, which means that any power level combinations are possible on consecutive uplink time slots in a given TDMA frame. For instance, for a GSM-900 small MS, the power control level ranges from +5 dBm to +33 dBm, which means that a high dynamic range is to be achieved in between two time slots by such an MS.

An important point to take into account is the possibility for a multislot MS with several slots allocated on the uplink to transmit an NB and an AB consecutively, or the opposite. Also, it may happen for a MS to transmit several ABs in a row in the same TDMA frame. Indeed, the AB may be used for PACKET CONTROL ACKNOWLEDGMENT message (see Chapter 5).

The other RF transmitter specifications (spectrum due to modulation, phase error, and spurious emissions) are kept as for GSM, and they also apply to multislot devices.

4.3.7 MS Synthesizer System Constraint

Within the RF transceiver system, the synthesizer is the device that is able to deliver an RF channel frequency with a very accurate definition. This section provides a summary of the system constraints that will be taken into account when designing an RF GPRS synthesizer. For a detailed overview on synthesis design, refer to [6].

4.3.7.1 Frequency Precision

As for the GSM voice service, the GPRS requirement on the RF channel accuracy is 0.1 ppm of the channel frequency for the MS, or 0.1 ppm compared with the signals that are received from the BTS. The only exception is for GSM-400, where the requirement is 0.2 ppm.

For example, in GSM-900, for the MS transmit carrier 912.4 MHz, the error on the mobile transmit frequency will be lower than 91.24 Hz.

4.3.7.2 Phase Noise Performance

Synthesizers are based on phase locked loops, and as such they are subject to phase noise, or jitter. In the frequency domain, this means that the frequency that is delivered by a synthesizer is not a pure sine wave. Rather, the power spectral density of a synthesizer exhibits some noise. The phase noise associated with the LO has an impact on the system specifications on both the transmitter and receiver parts of the RF transceiver.

On the transmitter side, the spectrum due to modulation, as well as the phase error requirements, are impacted. Indeed, the phase noise is directly added to the phase modulated signal, and therefore may degrade the modulation accuracy. On the receive section, the phase noise is an impairment that may impact the receiver performance (BER, BLER, and so on). It is usually taken into account in the IMs, in the estimation of the receiver sensitivity level.

4.3.7.3 Lock Time Performance

The mobile multislot classes were defined in Section 4.2.1. To summarize, we saw that two types of MS exist: type 1 (classes 1 to 12 and 19 to 29) are not able to receive and transmit at the same time, whereas type 2 (classes 13 to 18) have this capability. The different cases imply different constraints on the RF architecture, and in particular on the synthesizer. The synthesizer is responsible for the generation of the LO, used for the downconversion of the RF signal to baseband, and upconversion of the baseband signal to RF. An MS contains one or several synthesizers, depending on the architecture of the RF transceiver and on the supported multislot class. The multislot class of the MS has a huge impact on the synthesizer lock time performance.

The constraints that will be taken into account to derive the lock time of the synthesizers are:

- The (T_{ra}, T_{tb}) and (T_{rb}, T_{ta}) values that characterize the MS multislot class;
- The maximum TA value that can be commanded to the MS, since it reduces the delay available prior to a transmission;
- The size of the monitoring window that is necessary for a sufficient precision on the beacon channels (the BCCH carriers of the serving and neighbor cells) received signal measurements;

- The delay required to command the transmit, receive, and monitoring windows to the RF transceiver and to the synthesizer—this delay comprises the transmission of parameters such as the desired ARFC and the receiver gains.

The first two constraints listed above are inherent in the GPRS system definition, whereas the last two are strongly related to the implementation that is chosen in the MS. Indeed, the size of the monitoring window can vary according to the number of samples needed to achieve the desired accuracy. This window can vary in size, in the order of one time slot down to one-third of a time slot. The reduction of the monitoring window size may have an impact on the number of measurements that are required on the same carrier frequency.

For the sake of example, we will discuss the constraints related to class 12. The constraints for the other classes can easily be derived from this example. From Figure 4.32, we see that at least one time slot is available between the reception and the transmission slot, for the case in which two time slots

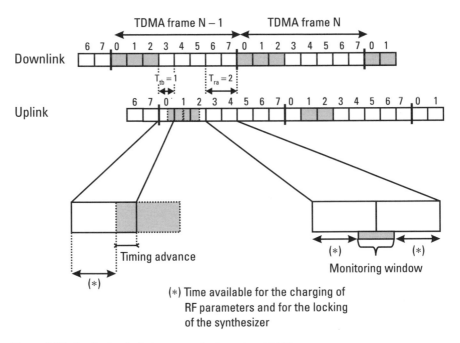

Figure 4.32 Synthesizer lock-time constraint for a class 12 MS.

are allocated on the uplink and three on the downlink. Also, at least two time slots are free between the end of the transmission and the start of the new reception slot, but these two time slots are also used to place a monitoring window.

We can see that between the receive and transmit time slots, one time slot minus the maximum TA is the minimum amount of time left for the RF parameter setting and for the synthesizer lock-up. The maximum TA value is 63 symbol periods, or 232.615 µs, so this represents 576.923 − 232.615 = 344.308 µs.

Between the transmit time slot of frame N and the next receive time slot in frame $N + 1$, two slots are available. During this period, the MS performs a received signal strength measurement (monitoring). This means that it must perform a switch from the transmit RF channel to the monitored RF channel, make the received signal measurement, and then switch from the monitored frequency to the receive RF channel in the next TDMA frame. In order to optimize the required synthesizer lock time, the monitoring window should be placed in the middle of the available period. If the duration of the monitoring window is T µs, the time left for the RF programming and synthesizer lock time is $(2 \times 576.923 - T)/2$ µs.

References

[1]   3GPP TS 05.05 Radio Transmission and Reception (R99).

[2]   Viterbi, A. J., "Error Bounds for Convolutional Codes and an Asymptotically Optimum Decoding Algorithm," *IEEE Trans. on Information Theory*, Vol. IT-13, April 1967, pp. 260–269.

[3]   Razavi, B., *RF Microelectronics*, Upper Saddle River, NJ: Prentice Hall, 1998.

[4]   Gray, P. R., and R. G. Meyer, "Future Directions in Silicon IC's for RF Personal Communications," *Proc. IEEE Custom Integrated Circuits Conference*, May 1995.

[5]   3GPP TS 05.50 Background for Radio Frequency (RF) Requirements (R99).

[6]   Crawford, J. A., *Frequency Synthesizer Design Handbook*, Norwood, MA: Artech House, 1994.

Selected Bibliography

3GPP TS 03.64 Overall Description of the GPRS Radio Interface; Stage 2 (R99).

3GPP TS 05.01 Physical Layer on the Radio Path; General Description (R99).

3GPP TS 05.02 Multiplexing and Multiple Access on the Radio Path (R99).

3GPP TS 05.03 Channel Coding (R99).

3GPP TS 05.04 Modulation (R99).

3GPP TS 05.08 Radio Subsystem Link Control (R99).

3GPP TS 05.10 Radio Subsystem Synchronization (R99).

5

Radio Interface: RLC/MAC Layer

This chapter provides a detailed description of the RLC/MAC layer procedures. The RLC/MAC layer is dedicated to the management of radio resources. As described in Chapter 3, in GPRS, temporary radio resources are allocated to the mobile in the downlink or uplink direction for the duration of the data transfer. MAC manages all the signaling necessary for their allocation (i.e., TBF establishment) and their release. It is also responsible for the mapping and multiplexing of signaling and data onto the different logical subchannels. The RLC protocol is in charge of the data transfer management during the TBF. It provides an acknowledged mode allowing selective retransmission of radio blocks as well as an unacknowledged mode of transmission.

For GPRS, the movement of the mobile through the network is entirely managed by a procedure of cell reselection. The mobile can perform autonomous cell reselection or reselection can be controlled by the network. In any case, the RLC/MAC layer manages this process.

The first section of this chapter introduces the RLC/MAC block structure. The RLC/MAC block is the most frequently used transport element on the air interface for signaling and data transfer between the mobile and the BSS. In order to be able to access the network, the mobile must acquire cell and network configuration parameters on the broadcast channel. The second section of the chapter describes how the network schedules these parameters and how the mobile monitors them. The cell reselection process, which partly relies on these broadcast parameters, is then explained. The three next sections deal with the management of the TBF. The way the mobile moni-

175

tors its PCH in order to detect a downlink transfer is described. Then the different procedures for uplink and downlink resource allocation, release, and data transfer (RLC protocol) are introduced.

The last section provides case studies dealing with RLC/MAC implementation aspects.

5.1 RLC/MAC Block Structure

The RLC/MAC block is the basic transport unit on the air interface that is used between the mobile and the network. It is used to carry data and RLC/MAC signaling.

In the previous chapter, the structure of the 52-multiframe was presented, and the concept of radio blocks was introduced. A radio block is defined as an information block transmitted over four consecutive bursts on four TDMA frames on a given PDCH.

One RLC data block is mapped onto one radio block, which is always transmitted on a packet data subchannel (PDTCH). One RLC/MAC control block is transmitted into one radio block on a signaling subchannel (PACCH, PCCCH, PBCCH).

The RLC/MAC control block is used to transmit RLC/MAC control messages, whereas the RLC data block contains data. A MAC header is added at the beginning of each type of radio block. A *block check sequence* (BCS) for error control detection is added at the end of the radio block.

The block formats are presented in the following section, with a brief definition of the different fields. The use of these fields for the GPRS layers will be studied in detail later in the chapter.

5.1.1 Control Block

The RLC/MAC block used for the transfer of control messages consists of a MAC header and an RLC/MAC control block, as shown in Table 5.1.

Table 5.1
RLC/MAC Block Structure for Control Messages

MAC header
RLC/MAC control block

RLC/MAC blocks used for control are encoded using the coding scheme CS-1. The size of the RLC/MAC control block is 22 bytes; the size of the MAC header is 1 byte.

5.1.1.1   Downlink RLC/MAC Control Block

Table 5.2 shows the format of the RLC/MAC control block with its MAC header for the downlink direction. The RLC/MAC control block consists of a control message contents field and an optional control header.

The MAC header contains the following elements:

- PT. This indicates whether the block is a control block or a data block.

- USF. This is used as an uplink multiplexing means when dynamic or extended dynamic allocations are used. It authorizes or refuses a transmission in the next block or set of four blocks following that received by a mobile. A number of mobiles can thus share a given uplink PDCH, but a single mobile transmits on one block at a given time. When resources are allocated, a given USF is reserved for a mobile on a given PDCH. The USF field can be set to the value FREE in a block on a PDCH that would support PCCCHs. FREE is used to indicate the presence of a PRACH, thereby allowing all mobiles in packet idle mode to send access requests on the PRACH.

- *Relative reserved block period* (RRBP). This indicates the number of frames that the mobile must wait before transmitting an RLC/MAC control block.

Table 5.2
Downlink RLC/MAC Control Block Format with MAC Header

8	7	6	5	4	3	2	1	
PT		RRBP		S/P		USF		MAC header
RBSN		RTI				FS	AC	Optional bytes
PR		TFI					D	
Control message content								

- *Supplementary/polling* (S/P). This indicates whether the RRBP field is valid.

The RLC/MAC header contains the following elements:

- *Reduced block sequence number* (RBSN), which gives the sequence number of the RLC/MAC control block;
- *Radio transaction identifier* (RTI), which is used to identify an RLC/MAC control message that has been segmented into two RLC/MAC control blocks;
- *Final segment* (FS), which indicates whether the RLC/MAC control block contains the FS of the segmented RLC/MAC control message;
- *Address control* (AC), which indicates the presence of an optional byte containing the PR, TFI, and D fields;
- *Temporary flow identifier* (TFI), which identifies a downlink or uplink TBF;
- *Direction* (D), which indicates the direction of the TBF identified by the TFI field;
- PR, which indicates the power reduction that has been used by the BTS to transmit the current block.

As its name indicates, the control message contents field contains an RLC/MAC control message.

5.1.1.2 Uplink RLC/MAC Control Block

Table 5.3 shows the format of the RLC/MAC control block for the uplink with its MAC header. The RLC/MAC control block consists of a control message contents field.

Table 5.3
Uplink RLC/MAC Control Block Format

8	7	6	5	4	3	2	1	
PT		Spare bits					R	MAC header
Control message content								

Table 5.4

RLC/MAC Block Structure for Data Transfer

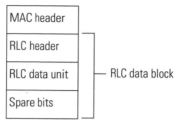

The MAC header contains:

- PT, which indicates the type of data within the block;
- R (retry), which indicates whether the mobile transmitted the access request message one time or more than one time during its most recent channel access.

5.1.2 RLC Data Block

The RLC/MAC block that is used for data transfer consists of a MAC header and an RLC data block. The RLC data block consists of an RLC header, an RLC data unit, and spare bits, as described in Table 5.4.

The size of the RLC block depends on the type of channel coding (CS-1, CS-2, CS-3, CS-4). A block can contain 184, 271, 315, or 431 bits, including the MAC header, and the number of spare bits is 0, 7, 3, 7 for CS1, CS2, CS3, and CS4 channel coding, respectively. The spare bits are set to 0 by the sending entity and ignored by the receiving entity.

Table 5.5

RLC Data Block Size by Coding Scheme

Coding Scheme	RLC Data Block Size Without Spare Bits (Bytes)	Number of Spare Bits
CS-1	22	0
CS-2	32	7
CS-3	38	3
CS-4	52	7

The size of the RLC data block for each of the channel coding schemes is shown in Table 5.5.

5.1.2.1 Downlink RLC Data Block

Table 5.6 shows the format of the RLC data block with its MAC header for the downlink data transfer.

The MAC header, which is common to the RLC data block and RLC/MAC control block for the downlink, was described in Section 5.1.1.1.

The RLC header contains the following fields:

- The PR indicates the power reduction used by the BTS to transmit the current block.

- The *temporary flow identifier* (TFI) identifies the ownership of the block. When resources are allocated, the TFI is used to identify the TBF.

- The *final block indicator* (FBI) indicates whether or not the RLC data block received is the last of the TBF.

- The *block sequence number* (BSN) is the sequence number of the RLC block in the TBF.

- *Extension* (E) indicates the presence of an optional byte.

Table 5.6
Downlink RLC Data Block with MAC Header

8	7	6	5	4	3	2	1	
PT		RRBP		S/P		USF		MAC header
PR		TFI					FBI	
BSN							E	
LI						M	E	
· · ·								Optional bytes
LI						M	E	
RLC Data								

- *More* (M) indicates the presence of another LLC frame in the data unit part.

- The *length indicator* (LI) makes it possible to delimit LLC frames within an RLC data block by giving the length of the data in the RLC data block belonging to an LLC frame. If this field is set several times, it indicates the length of the other LLC frames.

- The RLC data field can contain bytes for one or more LLC frames.

5.1.2.2 Uplink RLC Data Block

Table 5.7 shows the format of the RLC block for the uplink data transfer.

The MAC header does not contain exactly the same fields for the uplink as for the downlink. It contains the following fields:

- PT indicates whether the block transmitted is of the RLC data block type or the RLC/MAC control block type.

- *Countdown value* (CV) gives the number of RLC blocks associated with a TBF remaining to be transmitted.

- *Stall indicator* (SI) indicates an acknowledgment request from the mobile when the RLC protocol is stalled.

- R indicates whether the mobile transmitted the access request message one time or more than one time during its most recent channel access.

Table 5.7
RLC Data Block for Uplink with MAC Header

8	7	6	5	4	3	2	1	
PT		CV				SI	R	MAC header
Spare	PI	TFI					TI	
BSN							E	
LI					M		E	
TLLI								Optional bytes
PFI							E	
RLC Data								

The other fields are identical to the downlink RLC data block, with the exception of the following:

- *Temporary logical link identity* (TLLI), which identifies a GPRS user;
- *TLLI indicator* (TI), which indicates the presence of the TLLI field;
- *PFI indicator* (PI), which indicates the presence of the optional byte that contains the PFI;
- PFI, which identifies the packet flow context.

The E bit after the TFI field will be used in the future to indicate the presence of an additional byte due to an evolution of the protocol.

If the number of bytes corresponding to one or more LLC frames in the RLC data unit field cannot fill the RLC frame completely, it is filled with spare bits.

5.2   Broadcast Information Management

In each cell, the BCCH and PBCCH, if present, continuously broadcast information on the serving cell and neighbor cells' configuration.

The serving cell and neighbor cell parameters are broadcast within messages called SI messages on BCCH and *packet system information* (PSI) messages on PBCCH. Based on this information the MS is able to decide whether and how it may gain access to the system via the cell it is camping on.

The first subsection describes how the network schedules the different SI and PSI messages on BCCH and PBCCH, respectively. The next subsection details the types of parameters that can be found by the mobile within the different instances of these messages. Among them, cell reselection and frequency parameters are broadcast. Some details on the acquisition of cell reselection parameters are given, since their monitoring is dependent on whether they are sent on BCCH or PBCCH.

During a TBF establishment, the mobile is allocated frequency channels. It derives its frequency allocation from frequency parameters that are broadcast by the network. The last subsection explains how this is performed.

5.2.1   SI Message Scheduling

All information broadcast on the BCCH or PBCCH is carried in SI messages or PSI messages, respectively. As there are several SI and PSI message types having a different content, it is useful for the MS to know where these types of messages are located in a group of multiframes. A variable TC was introduced in the GSM standard to facilitate the MS operation. Each type of SI or PSI message is associated with a TC value.

For SI message types broadcast on the BCCH, TC is defined by the following formula:

$$TC = (FN \ DIV \ 51) \bmod 8 \tag{5.1}$$

This means that a given SI message type is located on a specific 51-multiframe identified by a 51-multiframe number equal to "FN DIV 51" every eight 51-multiframes (DIV means "divided by"). All SI message types are associated with a fixed TC value. For example, all the SI type 1 messages are sent when TC value is equal to 0. This means that all SI message types are scheduled in the same location within a group of 51-multiframes by any network manufacturer irrespective of operator.

On the PBCCH, the occurrence of PSI1 message type is defined by the following formula:

$$TC = (FN \ DIV \ 52) \bmod PSI1\_REPEAT\_PERIOD$$
$$\text{with } PSI1\_REPEAT\_PERIOD \tag{5.2}$$
$$\text{between 1 and 16}$$

A PSI1 message is broadcast by the network every PSI1_REPEAT_PERIOD occurrence of the 52-multiframe when TC is equal to 0. Unlike with the SI message type, the other PSI message types are not associated with a defined TC value fixed by the standard. This means there is an acquisition phase for the MS to determine the association between the PSI message type and the TC value when the MS camps on a new cell. The PSI message types other than the PSI1 message are divided into two categories:

- *High-repetition-rate PSIs.* These PSIs are broadcast on the PBCCH occurrence that is not used by the PSI1 in a sequence determined by the network. This sequence is repeated at each TC that is equal to 0 every PSI1_REPEAT_PERIOD of 52-multiframes. The PSI_COUNT_HR parameter broadcast by the network indicates the number of PSIs with high-repetition-rates.

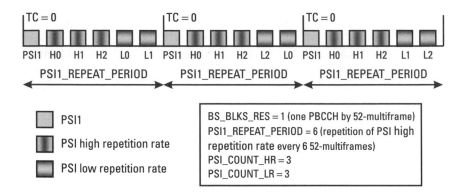

Figure 5.1 Mapping of PSI messages.

- *Low-repetition-rate PSIs.* These PSIs are broadcast on the PBCCH occurrences that are not used by the PSI1 and high-repetition-rate PSIs in a sequence determined by the network, and are repeated continuously. The sequence of these PSIs is repeated at the beginning of the hyperframe when the FN is equal to 0. The PSI_COUNT_LR parameter broadcast by the network indicates the number of PSIs with low-repetition-rates.

Note that as the PSI1_REPEAT_PERIOD, PSI_COUNT_HR, PSI_COUNT_LR parameters are broadcast in a PSI1 message, the MS needs to first read the PSI1 message to understand the mapping of the PSI message structure. Figure 5.1 gives an example of the PSI messages mapping in a given configuration.

5.2.2   MS Acquisition of Broadcast Information

5.2.2.1   Monitoring of BCCH Information

The BCCH is used to broadcast both GSM and GPRS network parameters. These parameters are the frequencies that are used in the cell, the neighbor cell frequencies, the GSM and GPRS logical channel description, and the access control parameters. The mobile uses the broadcast serving cell frequencies to derive its frequency allocation during resource assignment. It uses the neighbor cell frequencies for measurement and cell reselection purposes. The logical channel description indicates how the different logical channels are multiplexed on the time slots. The network broadcasts access control parameters and puts constraints on the access channels in order to avoid congestion.

These parameters are broadcast on the following SI messages:

- SI type 1 messages, which contain the serving cell frequency parameters;

- SI type 2, SI type 2bis, and SI type 2ter messages, which contain neighbor BCCH frequency list and access control parameters;

- SI type 3 messages, which contain control channel descriptions, cell options, and cell selection parameters;

- SI type 4 messages, which contain cell selection parameters and CBCH configuration;

- SI type 13 messages, which contains GPRS cell options.

SI type 2, 3, and 4 messages are always broadcast. SI type 1, 2bis, 2ter, and 13 messages are optional. Other SI messages exist but are not directly linked to the GPRS service.

Broadcast of the SI type 13 message in the cell indicates the availability of the GPRS service. The position of the SI type 13 message is indicated in either the SI type 3 or SI type 4 message. If PBCCH is not present in the cell, the SI type 13 message gives the GPRS cell parameters such as cell reselection mode, power control parameters, and the AB format to be used on the GPRS channels.

If PBCCH is present in the cell, the SI type 13 message indicates the description of the PBCCH (position, radio channel description, training sequence code, and PR of transmission compared to BCCH). The MS attempts to decode the full BCCH data of the serving cell at least every 30 seconds in order to detect any change in the network configuration. It also decodes the neighbor BCCH data block that contains the parameters affecting the cell reselection for each of the six strongest neighbor cells at least every 5 minutes.

5.2.2.2 Monitoring of PBCCH Information

The PBCCH broadcasts parameters relevant to both GSM and GPRS. When the PBCCH is present in the cell, it will be monitored by MSs that want to get GPRS attached and access to GPRS services. A GPRS mobile that monitors the PBCCH does not have to monitor the BCCH, since the BCCH information is also broadcast on PBCCH.

On PBCCH, the following types of messages are broadcast:

- PSI type 1 messages, which contain GPRS cell options, PCCCH description, and PRACH control parameters;
- PSI type 2 messages, which contain frequency parameters;
- PSI type 3 and PSI type 3bis messages, which contain the necessary cell reselection parameters of the serving cell and neighbor cells, and neighbor cell BCCH frequency information.

Other PSI messages are sent optionally by the network.

The mobile attempts to regularly decode the PSI type 1 messages of the serving cell (at least every 30 seconds). Within the PSI type 1 message, the network indicates whether a change took place within one or more types of broadcast PSI messages with the PBCCH_CHANGE_MARK parameter. The network may indicate which family of PSI messages has changed by using the PSI_CHANGE_FIELD parameter in order to avoid having the mobile rescan the complete PBCCH. Whenever a change in the PBCCH_CHANGE_MARK value is detected, the MS must reread the new PSI messages within the next 10 seconds.

Note that the MS may suspend its data transfer in order to decode the above information. However, in order to avoid data transfer interruption during packet transfer mode, the network can broadcast PSI on the PACCH. The mobile thus avoids leaving the data transfer for the monitoring of the PBCCH.

5.2.2.3 Cell Reselection Parameter Acquisition

For cell-reselection purposes, the mobile must acquire special parameters that are used for the evaluation of the cell-reselection criteria. Some are linked to the serving cell and others dedicated to the neighbor cells.

When there is no PBCCH in the serving cell, these parameters are broadcast by the network in the SI type 3 or SI type 4 messages. The serving cell parameters used for cell reselection are broadcast on the serving BCCH, whereas the neighbor cell parameters are broadcast on the neighbor BCCHs. The mobile must decode the SI3 and SI4 messages on the six strongest (in terms of RXLEV) neighbors' BCCH in order to acquire these parameters.

When the PBCCH is present in the serving cell, the network broadcasts both the neighbor cell and serving cell parameters that are used for cell reselection on this channel. They can be found in the PSI type 3 messages. The network may have to broadcast these parameters in several instances of the PSI3 message depending on the number of neighbor cells. Different encoding mechanisms have been introduced in order to provide a maximum number of parameters with a minimum number of PSI3 message instances. The acquisition time of neighbor cell parameters on PBCCH is much faster, since they are all transmitted on the same channel.

5.2.3 Frequency Parameters

When the network allocates radio resources to the mobile, it provides frequency parameters needed by the mobile to know its frequency channel allocation. The radio channel description used by the mobile consists of:

- One frequency in case of a nonhopping radio frequency channel;
- A list of frequencies, a MAIO, and a HSN (see Section 1.5.6.3) when frequency hopping is used.

The list of frequencies that is used by the mobile for GPRS transfer is called GPRS mobile allocation. The radio channel description is provided to the mobile during radio resource assignment.

When frequency hopping is used in the cell, the size of the mobile allocation can be very large. In order to avoid sending the list of frequencies directly during the assignment phase, special mechanisms have been defined.

5.2.3.1 Cell Allocation

The operator allocates a set of radio frequency channels to each cell. This set is defined as the *cell allocation* (CA). The CA defines the list of frequencies that can be assigned to the MS in the cell. In reality it defines the radio frequency channels that can be used during the period immediately following the very first resource assignment. Other radio frequency channels can be assigned to the mobile later, during the assignment phase. The CA is provided in the SI type 1 message or in the PSI type 2 messages when there is a PBCCH in the cell.

In the SI type 1 message, the CA is directly defined using special ARFCN encoding that depends on the number of frequencies provided.

In the PSI type 2 message, the network provides the mobile with *reference frequency lists* (RFLs). One RFL consists of a list of frequencies that are used in the cell. Up to four RFLs can be stored by the mobile. The CA is then defined in this case as the union of RFLs.

5.2.3.2 GPRS Mobile Allocation

The *GPRS mobile allocation* (GPRS MA) is defined as a subset of either the CA or RFLs. It is deduced from the CA or RFLs using a bitmap as illustrated in Figure 5.2.

Different encoding mechanisms have been defined in order to provide frequency lists within RFLs in PSI type 2 messages or CA in SI type 1 messages. For GSM-900, the number of frequencies is limited to 124. The encoding mechanism resides in a bitmap (each bit of the bitmap standing for an ARFCN from 1 to 124) indicating the frequencies of the list.

The MS must store up to seven GPRS MAs that can be broadcast in PSI type 2 messages. A GPRS MA within the PSI type 2 message is defined using a bitmap referring to one or more RFLs.

During the assignment, the network can provide the radio channel description with either:

- One frequency in case of a nonhopping radio frequency channel;
- One GPRS MA number, one HSN, and one MAIO (referred to as "indirect encoding" in the GSM specifications);
- A bitmap referring to one or more RFLs (broadcast on PBCCH) of the CA (broadcast on BCCH or PBCCH), one HSN, and one MAIO (referred to as "direct encoding 1" in the GSM specifications);
- A list of frequencies, one MAIO, and one HSN (referred to as "direct encoding 2" in the GSM specifications).

Note that the drawback of the direct encodings 1 and 2 is the large amount of space required in the assignment message. If many frequencies are

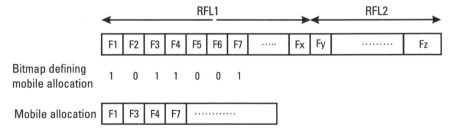

Figure 5.2 Example of mobile allocation definition.

sent, it will be necessary to provide the assignment message in two radio blocks. This increases the delay and complexity of the establishment. The drawback of the indirect encoding is that more space is required in the PSI2 messages to broadcast the RFLs, leading to a lower repetition rate of some other PSI messages.

5.3 Cell Reselection

The cell reselection can be controlled either autonomously by the mobile or by the network. It is based on measurements performed by the mobile. The network can order that these measurements be reported periodically.

Three cell reselection modes NC0, NC1, and NC2 have been defined:

- *NC0.* In this mode, the GPRS mobile performs autonomous cell reselection without sending measurement reports to the network.
- *NC1.* In this mode, the GPRS mobile performs autonomous cell reselection and periodically sends measurement reports to the network.
- *NC2.* In this mode, the network controls the cell reselection. The mobile sends measurement reports to the network.

When the mobile performs autonomous cell reselection, it chooses a new cell and triggers a cell reselection on its own. The NC2 mode allows the network to control the mobility of GPRS users within the network.

The GPRS cell reselection mode for a GPRS-attached MS is given by the network control mode (NETWORK_CONTROL_ORDER parameter), which is broadcast on the BCCH or PBCCH. The mobile behavior is determined by both its GMM state and the network control mode. Whatever the value of the network control mode, when the MS is in GMM STANDBY state, it performs autonomous cell reselection and does not send measurement reports to the network. In GMM READY state, the mobile performs cell reselection according to the network control mode.

Two criteria are defined for autonomous cell reselection: one is based on the (C1, C2) criteria. It corresponds to the GSM cell-reselection criteria. The other, based on the (C'1, C31, C32) criteria, has been introduced for GPRS. All these criteria are based on *received signal level* (RXLEV) measurements in the serving cell and in the neighbor cells.

Table 5.8
Mode of Reselection and Criteria for Cell Reselection

NETWORK_ CONTROL_ ORDER Value	GMM State of the MS	Mode of Cell Reselection	Criteria If PBCCH Exists	Criteria If PBCCH Does Not Exist
NC0	Standby/Ready	Autonomous cell reselection	C'1, C31, C32	C1, C2*
NC1	Standby	Autonomous cell reselection	C'1, C31, C32	C1, C2*
	Ready	Autonomous cell reselection with measurement reports	C'1, C31, C32	C1, C2*
NC2	Standby	Autonomous cell reselection	C'1, C31, C32	C1, C2*
	Ready	Network-controlled cell reselection with measurement reports	—	—

* Except if the GPRS cell reselection parameters are sent to the MS in an RLC/MAC control message.

In a cell where PBCCH does not exist (GPRS cell reselection parameters are not broadcast), the MS performs cell reselection based on (C1, C2) criteria; except if the cell reselection parameters are sent explicitly to the mobile in one control message, in which case the mobile performs autonomous cell reselection based on criteria (C'1, C31, C32). In a cell where PBCCH exists, the MS performs cell reselection based on (C'1, C31, C32) criteria.

Table 5.8 summarizes the different reselection modes and the criteria used.

5.3.1 Measurements

When the mobile camps on a cell, it receives from the network the list of the neighbor cells identified by their BCCH frequencies and their BSICs. The mobile measures periodically the RXLEV on these neighbor BCCHs and checks the BSIC of the BCCH carriers.

These RXLEV measurements are used in autonomous cell reselection mode and reported to the network in case of cell reselection controlled by the network and NC1 mode.

5.3.1.1   In Packet Idle Mode

In packet idle mode, the mobile measures the RXLEV on the BCCH carrier of the serving cell and on each BCCH carrier of the neighboring cells. It calculates an average RXLEV on each carrier, with at least five measurements during a maximum period corresponding to the maximum of 5 seconds or five consecutive paging blocks of the mobile. The mobile must perform a minimum of one measurement on a cell every 4 seconds or a maximum of 20 measurements every second. The mobile maintains a list of the six strongest received BCCH carriers, which is updated every averaging period.

The mobile must decode the BSIC on the BCCHs of the six most powerful cells every 14 paging block periods or at least every 10 seconds.

5.3.1.2   In Packet Transfer Mode

In packet transfer mode, the mobile measures the RXLEV on the BCCH carrier of the serving cell and on each BCCH carrier of the neighboring cells. It calculates an average RXLEV on each carrier during a period of 5 seconds, with at least five measurements.

The mobile performs a measurement on each TDMA frame for at least one of the given BCCH carriers, one after the other. The mobile must decode the BSIC on the BCCHs of the six most powerful cells at least every 10 seconds. The mobile does this decoding during the two idle frames of the 52-multiframe.

5.3.2   Criteria

In this section, the criteria C1, C'1, C2, C31, and C32 that were introduced in the previous section are described in detail. The way these parameters are used in the cell reselection process will be described in the next section.

5.3.2.1   C1 Criterion

The path loss criterion parameter C1 is used as a minimum signal level criterion for cell reselection when there is no PBCCH in the cell.

C1 is defined by the following formula:

$$C1 = (A - \text{Max}[B,0]) \tag{5.3}$$

$$A = \text{RXLEV} - \text{RXLEV\_ACCESS\_MIN} \tag{5.4}$$

$$B = \text{MS\_TXPWR\_MAX\_CCH} - P \tag{5.5}$$

RXLEV_ACCESS_MIN represents the minimum RXLEV at the MS to access the cell. MS_TXPWR_MAX_CCH represents the maximum transmit power level allowed to the MS when accessing the cell. P is the maximum RF output power of the MS (parameter specific to the MS).

A represents the reception margin. The MS is allowed to access the cell if its RXLEV is higher than RXLEV_ACCESS_MIN; that is, $A > 0$. If $A > 0$, the mobile is in the cell coverage and the downlink is good enough. If $A < 0$, the mobile is outside the cell coverage.

B represents the mobile transmission capability margin. If $B < 0$, the transmission capabilities of the mobile are sufficient; if the mobile is in the cell coverage $A > 0$, the cell can be selected. If $B > 0$ and $A - B > 0$, the mobile transmission capabilities are compensated by the reception margin and the cell can be selected. If $B > 0$ and $A - B < 0$, the mobile transmission capabilities are not compensated, and the mobile cannot select the cell.

Thus the path loss criterion is satisfied if C1 > 0.

5.3.2.2   C'1 Criterion

The path loss criterion parameter C'1 is used as a minimum signal level criterion for GPRS cell reselection in the same way as for GSM idle mode. C'1 is the same as C1, except that GPRS-specific parameters are used instead.

C'1 is defined by the following formula:

$$\text{C'1} = (A - \text{Max}[B,0]) \tag{5.6}$$

where

$$A = \text{RXLEV} - \text{GPRS\_RXLEV\_ACCESS\_MIN} \tag{5.7}$$

and

$$B = \text{GPRS\_MS\_TXPWR\_MAX\_CCH} - P \tag{5.8}$$

5.3.2.3   C2 Criterion

The C2 criterion is used for cell ranking in the GSM cell-reselection process. It is computed with the following formulas:

If T < PENALTY_TIME
$$\text{C2} = \text{C1} + \text{CELL\_RESELECT\_OFFSET} - \text{TEMPORARY\_OFFSET} \tag{5.9}$$

If T > PENALTY_TIME
C2 = C1 + CELL_RESELECT_OFFSET   (5.10)

T is a timer that is started from 0 at the time the cell enters in the list of strongest carriers. CELL_RESELECT_OFFSET is a parameter that is used to prioritize one cell in relation to the others. TEMPORARY_OFFSET is a parameter used to penalize during PENALTY_TIME when this cell just enters the list of strongest carriers.

5.3.2.4 C31 Criterion

The signal level threshold criterion parameter C31 for hierarchical cell structures (HCS) is used to determine whether prioritized hierarchical GPRS cell reselection applies.

The C31 criterion allows cells for GPRS-attached mobiles to be prioritized during autonomous cell reselection. A GPRS-attached MS will preferably select the cell having the highest priority as indicated by the parameter PRIORITY_CLASS. However, a sufficient RXLEV in the cell (HCS_THR parameter) is required for it to belong within the highest-priority class. If the signal level becomes too low, a determination of lowest priority is made.

The C31 criterion contains a time-based offset. This offset can be used to penalize a cell, belonging to another priority level as the serving cell, during GPRS_PENALTY_TIME.

The C31 criteria for the serving and neighbor cells are evaluated with the following formula:

C31(serv) = RXLEV(serv) − HCS_THR(serv)   (5.11)

if PRIORITY_CLASS(neigh) = PRIORITY_CLASS (serv)
    C31(neigh) = RXLEV (neigh) − HCS_THR(neigh)   (5.12)

if PRIORITY_CLASS(neigh) ≠ PRIORITY_CLASS(serv)
    if T < GPRS_PENALTY_TIME

C31(neigh) = RXLEV(neigh)
    − HCS_THR(neigh) − GPRS_TEMPORARY_OFFSET   (5.13)

if T ≥ GPRS_PENALTY_TIME
C31(neigh) = RXLEV(neigh) − HCS_THR(neigh)   (5.14)

where HCS_THR is the signal threshold for applying HCS GPRS reselection, and GPRS_PENALTY_TIME gives the duration for which the temporary offset GPRS_TEMPORARY_OFFSET is applied. T is a timer that is started from 0 at the time the cell enters in the list of strongest carriers.

5.3.2.5   C32 Criterion

The cell-ranking criterion parameter (C32) is used to select cells among those with the same priority and is defined for the serving cell and the neighbor cells by:

$$C32(serv) = C'1(serv) \tag{5.15}$$

if PRIORITY_CLASS(neigh) = PRIORITY_CLASS(serv)
  if T < GPRS_PENALTY_TIME
  C32(neigh) = C1(neigh) + GPRS_RESELECT_OFFSET(neigh)
    – GPRS_TEMPORARY_OFFSET   (5.16)

  if $T \geq$ GPRS_PENALTY_TIME
  C32(neigh) = C1(neigh) + GPRS_RESELECT_OFFSET(neigh)   (5.17)

if PRIORITY_CLASS(neigh) $\neq$(PRIORITY_CLASS(serv)
  C32(neigh) = C1(neigh) + GPRS_RESELECT_OFFSET(neigh)   (5.18)

The C32 criterion also contains a time-based offset that can be used to penalize a cell belonging to the same priority level during a certain time. It could be use to perform speed discrimination. The GPRS_RESELECT_OFFSET parameter allows a neighbor cell to be penalized.

When the MS is in GMM READY state (a transfer can be ongoing or about to start), in order to avoid unnecessary cell reselection or to be sure that it is really needed, an hysteresis GPRS_CELL_RESELECT_HYSTERESIS is subtracted from the C32 criteria of the neighbor cells. This hysteresis is also subtracted from the C31 criterion if requested by the network.

When the neighbor cell does not belong to the same RA as the serving cell, an hysteresis RA_RESELECT_HYSTERESIS is subtracted from the C32 criterion of the neighbor cell. This is done to avoid changing from one RA to the other and so triggering an RA update procedure, increasing the signaling within the network.

In case cell reselection occurred within the previous 15 seconds, all neighbor cells are penalized by 5 dB in order to avoid ping-pong between cells.

All the GPRS cell-reselection parameters described in this section are broadcast on the PBCCH carrier of the serving cell. The cell-reselection parameters used for calculation of the (C1, C2) criteria are broadcast on the BCCH carrier of the serving cell and the BCCH carriers of the neighbor cells.

5.3.3 Autonomous Cell-Reselection Process

The process of cell reselection is either based on the (C1, C2) criteria or (C'1, C31, C32) criteria. (C1, C2) criteria are used when there is no PBCCH in the serving cell and the cell-reselection parameters needed for (C'1, C31, C32) criteria evaluation have not been explicitly transmitted to the mobile.

(C'1, C31, C32) criteria are used when the PBCCH is transmitted in the serving cell, or when the cell reselection parameters needed for (C'1, C31, C32) criteria evaluation have been explicitly transmitted to the mobile and there is no PBCCH.

5.3.3.1 Procedure for (C1, C2)

The mobile must compute the C1 and C2 values for the serving cell and the neighbor cells at least every 5 seconds. A cell-reselection is triggered if one of the following events occurs:

- The path loss criterion parameter C1 indicates that the path loss in the serving cell has become too high (i.e., C1 < 0).
- Downlink signaling failure is experienced (see Section 5.4.5).
- The serving cell has become barred.
- A neighbor cell for which the C2 criterion is higher than the serving cell for a period of 5 seconds is detected. When the neighbor cell is not in the same RA or if the mobile is in GMM READY state, the C2 value for the neighbor cell must exceed the C2 value for the serving cell by at least CELL_RESELECT_HYSTERESYS. This allows for the avoidance of unnecessary cell reselection when the mobile is in READY state or between two different RAs. For a cell reselection occurring in the previous 15 seconds, the C2 value for the neighbor cell must exceed the C2 value for the serving cell by at least 5 dB for a period of 5 seconds.

- A random access attempt is still unsuccessful after "Max retrans" repetitions, Max retrans being a parameter broadcast on the control channel.

5.3.3.2   Procedure for (C'1, C31, C32)

The mobile computes the values of the C'1, C31, and C32 criteria for the serving cell and the nonserving cells at least every second or at every new measurement, whichever is the greatest. A cell reselection is then triggered if one of the following events occurs:

- The path loss criterion parameter C'1 indicates that the path loss to the cell has become too high (i.e., C'1 < 0).

- A nonserving suitable cell is evaluated to be better than the serving cell. The best cell is the cell with the highest C32 value among: (a) the cells that have the highest PRIORITY_CLASS among the cells that fulfill C31 >=0, or (b) all cells, if no cell fulfills the criterion C31 >= 0.

- Downlink signaling failure is experienced (see Section 5.4.5).

- The cell on which the mobile is camping has become barred.

- A   random   access   attempt   is   still   unsuccessful   after MAX_RETRANS_RATE attempt.

The cell that the mobile was camped on must not be returned to within 5 seconds.

5.3.4   **Measurement Report Sending**

When the cell reselection mode given by the NETWORK_CON-TROL_ORDER parameter indicates NC1 or NC2 and the MS is in GMM READY state, the MS reports measurements in the PACKET MEASURE-MENT REPORT message to the network. This message contains:

- The measured RXLEV for the serving cell;

- In packet idle mode, the average interference level γ_{ch} for the serving cell;

- The measured RXLEV from the nonserving cells.

The reporting period is defined as follows:

- In packet idle mode, the reporting period is indicated by NC_REPORTING_PERIOD_I parameter;
- In packet transfer mode, the reporting period is indicated by the NC_REPORTING_PERIOD_T parameter.

NC_REPORTING_PERIOD_I and NC_REPORTING_PERIOD_T parameters are broadcast on PBCCH. They can also be broadcast on PACCH in case there is no PBCCH.

5.3.5 Network-Controlled Cell-Reselection Process

When the cell-reselection mode indicates NC2 and the mobile is in GMM READY state, the network controls the cell-reselection process. Based on the measurements reported by the mobile (RXLEV and RXQUAL only during packet transfer) or by the BTS during packet transfer mode, the network decides whether a cell reselection must be triggered.

When the decision to trigger a cell reselection is made by the network, it sends a PACKET CELL CHANGE ORDER message. When the MS receives the command, it reselects the cell according to the included cell description (BCCH frequency and BSIC) and changes the network control mode according to the NETWORK_CONTROL_ORDER parameter included in the command.

Note that in order to reduce the cell-reselection delay, the MS may continue its transfer in the old cell until it decodes the BCCH or PBCCH information necessary to reestablish a TBF in the new neighbor cell.

5.4 Listening to MS Paging Blocks

If it needs to send GPRS data toward a given MS, the network analyzes the MS GMM state in order to decide whether it may or not initiate a paging procedure. When the MS is in GMM STANDBY state, the network does not know accurately the GPRS MS location. In fact, if the mobile camps on a new cell in (packet) idle mode within the same RA, it does not notify the network of a cell change in GMM STANDBY state. The paging procedure allows for the location of an MS within a whole RA by sending a paging message in all cells belonging to this MS's RA. When the MS detects a paging that is intended for it while it is in GMM STANDBY state, it answers by sending any LLC frame and changes into GMM READY state. Upon receipt of an LLC frame sent as a paging response, the network is able to send a resource assignment command for a downlink TBF.

Note   The network does not need to initiate a paging procedure if the MS is in GMM READY state because in this state the MS notifies the network of each cell change in a given RA. Therefore, the network is able to send directly a resource assignment command for a downlink TBF toward an MS in GMM READY state.

5.4.1   Network Operating Modes

Table 5.9 summarizes the MS behavior for the reception of paging messages for circuit and GPRS services as a function of the network-operating mode, MS class, RR operating mode (idle mode, dedicated mode, transfer mode), and presence or absence of PCCCH channels.

Note that in a network mode III that supports PCCCH channels, a class B MS will ensure a double idle mode operation; actually, the MS must monitor CCCH and PCCCH channels. As this behavior has an impact on the MS battery autonomy, some mobiles will be configured in class C in a network-operating mode III.

Table 5.9
MS Behavior for Receipt of Paging Message

	Class A	Class B	Class C
Mode I with PCCCH channels	*(Packet) idle mode:* decoding of PCCCH channels for CS and GPRS incoming calls.	*(Packet) idle mode:* decoding of PCCCH channels for CS and GPRS incoming calls.	*Class C configured in CS mode:* decoding of CCCH channels to detect CS incoming calls during idle mode.
	Dedicated mode: decoding of PCCCH channels to detect GPRS incoming calls.	*Dedicated mode:* no decoding of PCCCH channels to detect GPRS incoming calls.	*Class C configured in GPRS mode:* decoding of PCCCH channels to detect GPRS incoming calls during packet idle mode.
	Packet transfer mode: decoding of PACCH to detect CS incoming calls.	*Packet transfer mode:* decoding of PACCH to detect CS incoming calls.	

Table 5.9 (continued)

	Class A	Class B	Class C
Mode I without PCCCH channels	*(Packet) idle mode:* decoding of CCCH channels for CS and GPRS incoming calls. *Dedicated mode:* decoding of CCCH channels to detect GPRS incoming calls. Packet transfer mode: decoding of PACCH to detect CS incoming calls.	*(Packet) idle mode:* decoding of CCCH channels for CS and GPRS incoming calls. *Dedicated mode:* no decoding of CCCH channels to detect GPRS incoming calls. Packet transfer mode: decoding of PACCH to detect CS incoming calls.	*Class C configured in CS mode:* decoding of CCCH channels to detect CS incoming calls during idle mode. *Class C configured in GPRS mode:* decoding of CCCH channels to detect GPRS incoming calls during packet idle mode.
Mode II	*(Packet) idle mode:* decoding of CCCH channels for CS and GPRS incoming calls. *Dedicated mode:* decoding of CCCH channels to detect GPRS incoming calls. *Packet transfer mode:* decoding of CCCH to detect CS incoming calls.	*(Packet) idle mode:* decoding of CCCH channels for CS and GPRS incoming calls. *Dedicated mode:* no decoding of CCCH channels to detect GPRS incoming calls. *Packet transfer mode:* no obligation to decode CCCH for CS incoming-call detection.	*Class C configured in CS mode:* decoding of CCCH channels to detect CS incoming calls during idle mode. *Class C configured in GPRS mode:* decoding of CCCH channels to detect GPRS incoming calls during packet idle mode.
Mode III with PCCCH channels	*(Packet) idle mode:* decoding of PCCCH channels for GPRS incoming calls and decoding of CCCH channels for CS incoming calls. *Dedicated mode:* decoding of PCCCH channels to detect GPRS incoming calls. *Packet transfer mode:* decoding of CCCH channels to detect CS incoming calls.	*(Packet) idle mode:* decoding of PCCCH channels for GPRS incoming calls and decoding of CCCH channels for CS incoming calls. *Dedicated mode:* no decoding of PCCCH channels to detect GPRS incoming calls. *Packet transfer mode:* no obligation to decode CCCH for CS incoming-call detection.	*Class C configured in CS mode:* decoding of CCCH channels to detect CS incoming calls during idle mode. *Class C configured in GPRS mode:* decoding of PCCCH channels to detect GPRS incoming calls during packet idle mode.

Table 5.9 (continued)

	Class A	Class B	Class C
Mode III without PCCCH channels	*(Packet) idle mode:* decoding of CCCH channels for CS and GPRS incoming calls. *Dedicated mode:* decoding of CCCH channels to detect GPRS incoming calls. *Packet transfer mode:* decoding of CCCH channels to detect CS incoming calls.	*(Packet) idle mode:* decoding of CCCH channels for CS and GPRS incoming calls. *Dedicated mode:* no decoding of CCCH channels to detect GPRS incoming calls. *Packet transfer mode:* no obligation to decode CCCH for CS incoming-call detection.	*Class C configured in CS mode:* decoding of CCCH channels to detect CS incoming calls during idle mode. *Class C configured in GPRS mode:* decoding of CCCH channels to detect GPRS incoming calls during packet idle mode.

5.4.2   DRX Mode

In order to minimize the power consumption, the MS in idle mode is not required to listen continuously to the amount of information provided by the network. The (P)PCH logical channels have been split into several paging subchannels; all paging messages addressed to an MS with a given IMSI are sent on the same paging subchannel. This feature is called DRX.

Note that the DRX feature allows for an increase in battery lifetime, but at the expense of a small increase in time delay for the establishment of circuit and GPRS incoming calls.

The method whereby paging subchannels are accessed is determined by either the network, in the case of PCH channels, or by the MS, in the case of PPCH channels.

5.4.2.1   (P)CCCH_GROUP Definition

As there may be multiple CCCH or PCCCH logical channels on different time slots of the same carrier or on different carriers in the same cell, the MS will first determine the right carrier and the right time slot for paging-channel decoding. Thus a CCCH_GROUP or PCCCH_GROUP addresses a specific CCCH or PCCCH for paging-message listening but also random accesses on RACH or PRACH.

The CCCH channel is found on time slot 0 of the BCCH carrier, but may also be found on time slots 2, 4, and 6. The parameter BS_CC_CHANS broadcast in the BCCH defines the number of CCCHs.

The MS is able to define the right time slot CCCH channel according to the parameter CCCH_GROUP. CCCH_GROUP is a number between 0 and BS_CC_CHANS − 1.The lowest-numbered CCCH_GROUP is mapped on the lowest-numbered time slot carrying CCCH, the next-higher-numbered CCCH_GROUP is mapped on the next-higher-numbered time slot carrying the CCCH, and so on. The CCCH_GROUP is defined by the following formula:

$$CCCH\_GROUP = \\ \left(\left(IMSI \bmod 1,000\right) \bmod \left(BS\_CC\_CHANS \times N\right)\right) div\ N \qquad (5.19)$$

with N = number of paging blocks "available" in a 51-multiframe on one CCCH × number of 51-multiframes between transmission of paging messages to MSs.

Note that the CCCH_GROUP is used by MSs in circuit mode or GPRS mode for all network-operating modes if PBCCH/PCCCHs are not present in the cell. The MS determines the N value from parameters broadcast on BCCH.

The PCCCH channel may be found on several carriers and on several time slots per carrier. The parameter BS_PCC_CHANS (maximum value: 16) defines the number of PDCHs carrying the PCCCHs. The MS is able to identify the specific PCCCH according to the PCCCH_GROUP parameter. The PCCCH_GROUP is numbered from 0 to BS_PCC_CHANS − 1. The network broadcasts to the MS the organization of PCCCH by stating the list of used carriers. The mapping between the PCCCH_GROUP and the physical channel follows the PCCCH description broadcast on PBCCH. The lowest-numbered PCCCH_GROUP is mapped on the lowest-numbered time slot carrying PCCCH on the first PCCCH carrier. The next-higher-numbered PCCCH_GROUP is mapped on the next-higher-numbered time slot carrying the same carrier, and so on. When all time slots of the first PCCCH carrier are used, the next-higher-numbered PCCCH_GROUP is mapped on the lowest-numbered time slot carrying PCCCH on the next PDCH that carries, and so on. The PCCCH_GROUP is defined by the following formula:

$$PCCCH\_GROUP = \\ (IMSI \bmod 1,000) \bmod BS\_PCC\_CHANS \qquad (5.20)$$

Note that the formula given above is applied with respect to GPRS-attached mobiles if PBCCH/PCCCHs are present in the cell. Nevertheless, if the PBCCH/PCCCHs are not present in the cell, the GPRS mobile may choose to apply the PCCCH_GROUP definition instead of the CCCH_GROUP definition. In this case, the formula given above is slightly modified. The choice between these two different procedures is negotiated during the attachment.

5.4.2.2 PAGING_GROUP Definition

As there are several paging subchannels in the same physical channel carrying the CCCH identified by a CCCH_GROUP or on the same PDCH carrying the PCCCH identified by a PCCCH_GROUP, the MS will determine the right paging subchannel for paging blocks decoding according to the parameter PAGING_GROUP. This paging subchannel is defined as a set of multiframes carrying the CCCH or the PCCCH. This set of multiframes defines the periodicity at which the MS is required to decode paging messages. This periodicity of listening is configured either by the network, when paging occurs on the physical channel carrying the CCCH, or by the MS, when paging occurs on the PDCH carrying the PCCCH or when paging occurs on the physical channel carrying the CCCH and the mobile applies the PCCCH_GROUP definition.

On CCCH Logical Channels

The BS_PA_MFRMS parameter defines the number of 51-multiframes for the periodicity of PCH subchannel paging decoding. The value is broadcast on BCCH and may range from 2 to 9. For instance, if the value is equal to 9, the MS will decode its paging subchannel every nine 51-multiframes. The number of paging subchannels N on one physical channel carrying the CCCH is equal to the number of paging blocks in one 51-multiframe multiplied by the periodicity of subchannel paging decoding (BS_PA_MFRMS value). The PAGING_GROUP is numbered from 0 to $N - 1$. The PAGING_GROUP is defined by the following formula:

$$PAGING\_GROUP\ (0...N-1) =$$
$$((IMSI\ \mathrm{mod}\ 1,000)\ \mathrm{mod}\ (BS\_CC\_CHANS \times N))\ \mathrm{mod}\ N \quad (5.21)$$

with N = number of paging blocks "available" on one CCCH (value deduced from parameters broadcast on BCCH).

In order to find the required 51-multiframe containing the paging subchannel among the cycle of BS_PA_MFRMS 51-multiframes, the following equality will be checked:

$$PAGING\_GROUP \ div \ (N \ div \ BS\_PA\_MFRMS) = \\ (FN \ div \ 51) \ mod \ (BS\_PA\_MFRMS) \tag{5.22}$$

with FN = frame number.

The required paging subchannel is identified in the required 51-multiframe by the following formula:

$$Paging \ block \ index \ = \ PAGING\_GROUP \ mod \\ (N \ div \ BS\_PA\_MFRMS) \tag{5.23}$$

Figure 5.3 shows an example of the use of CCCH_GROUP and PAGING_GROUP concepts for the required paging subchannel PCH decoding.

On PCCCH Logical Channels

The SPLIT_PG_CYCLE parameter defines the occurrence of paging blocks on the PDCH carrying the PCCCH belonging to the MS in DRX mode by specifying a number of subcycles used for PPCH subchannel paging decoding within a cycle of 64 52-multiframes. The SPLIT_PG_CYCLE value is fixed by the MS and is sent by the MS to the network during the GPRS attach procedure or during the RA update procedure.

The number of paging subchannels M used on every 64 52-multiframe is equal to a number of paging blocks in one 52-multiframe multiplied by 64, given by the following formula:

$$(12 - BS\_PAG\_BLKS\_RES - BS\_PBCCH\_BLKS) \times 64 \tag{5.24}$$

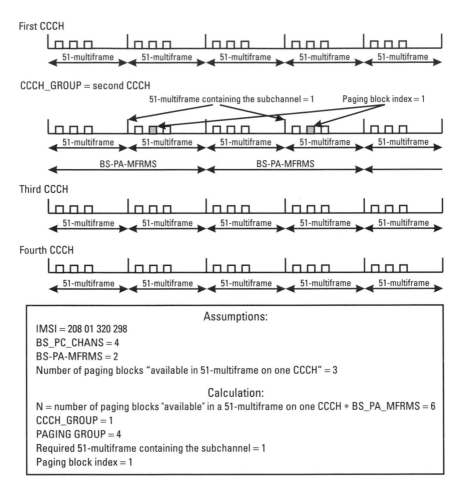

Figure 5.3   Example of the use of CCCH_GROUP and PAGING_GROUP concepts for the required paging subchannel PCH decoding.

The PAGING_GROUP is numbered from 0 to N − 1. The PAGING_GROUP is defined by the following formula:

$$
\begin{aligned}
&PAGING\_GROUP \\
&= \left(\begin{array}{l} (IMSI \bmod 1000)\, \mathrm{div}\; BS\_PCC\_CHANS \\ +IMSI \bmod 1000 \\ +Max\big((n\times M)\,\mathrm{div}\; SPLIT\_PG\_CYCLE, n\big) \end{array} \right) \bmod M
\end{aligned} \tag{5.25}
$$

with n: subblock number between 0 and MIN[M, SPLIT_PG_CYCLE].

The term MIN[M, SPLIT_PG_CYCLE] makes it possible to prevent more subperiods than paging blocks if SPLIT_PG_CYCLE > M.

The term Max (($n \times N$) div SPLIT_PG_CYCLE),n) makes it possible to verify that there are more subperiods than paging blocks and to have, for each new value of n, a value of the term MAX which is different from the ones previously calculated, ensuring that all the PAGING_GROUP values for each subcycle are different from one another.

Note that:

1. The PAGING_GROUP formula for PCCCH gives a number of values, each value being associated with a subcycle. The number of the PAGING_GROUP values is equal to the value MIN[M,SPLIT_PG_CYCLE]. The mobile will thus have to monitor the PCH a number of times equal to the value of the SPLIT_PG_CYCLE every 64 52-multiframes.

2. The PAGING_GROUP formula given above is slightly modified if a GPRS MS decides to apply SPLIT_PG_CYCLE on CCCH.

3. The term 64/SPLIT_PG_CYCLE is equivalent to the BS_PA_MFRMS.

4. The lower the BS_PA_MFRMS value (or the higher the SPLIT_PG_CYCLE value), the higher the recurrence of paging decoding and the lower the MS's autonomy in packet idle mode.

In order to find the required 52-multiframe containing the paging subchannel among the cycle of 64 52-multiframes, the following equality will be checked:

$$PAGING\_GROUP \ \mathrm{div} \left(M \ \mathrm{div} \ 64 \right)$$
$$= \left(FN \ \mathrm{div} \ 52 \right) \mathrm{mod} \ 64 \tag{5.26}$$

with FN = frame number.

The required paging subchannel is identified in the required 52-multiframe by the following formula:

$$PAGING\_GROUP \ \mathrm{mod} \left(M \ \mathrm{div} \ 64 \right) \tag{5.27}$$

Note that the same concept is used for the required paging subchannel decoding between PCH and PPCH. The main difference is bound by the number of PAGING_GROUP values depending on whether or not PCCCH logical channels are present. For one PAGING_GROUP value, the required paging subchannel PPCH is identified in the same way as the required PCH (see Figure 5.3).

5.4.3 Non-DRX Mode

After a TBF release or a measurement report in idle mode the mobile reverts to the non-DRX mode, in which it must decode all CCCH or PCCCH blocks, independently of its DRX period. If the network needs to initiate a new downlink TBF, it sends the downlink TBF allocation on any (P)CCCH related to the (P)CCCH_GROUP of the MS. The downlink TBF establishment is quicker for a MS in non-DRX mode because the network does not need to wait for the MS paging subchannel in order to send the paging message. After a TBF release, the non-DRX period is equal to the minimum value of two values. One value is set by the network (DRX_TIMER_MAX parameter) and given on broadcast channels, and the other one is set by the MS (NON_DRX_TIMER parameter) during the GPRS-attach procedure. At the end of a measurement report in idle mode, the non-DRX period is equal to a value set by the network (NC_NON_DRX_PERIOD parameter) and given on broadcast channels.

Note that:

1. There is a high probability that a new downlink TBF needs to be established a short time after the end of an uplink TBF or a downlink TBF. In the case of client-server relation, a downlink TBF may occur shortly after an uplink TBF. For instance, in the case where the uplink TBF is used to send a request, a downlink TBF may be established for the transfer of the response. There may be a latency period between two consecutive IP packets sent to a given MS. This means in some cases that several downlink TBFs are needed for the sending of consecutive IP packets toward the MS.

2. The non-DRX mode has an impact on MS autonomy in idle mode because in this period the MS must continuously decode the CCCH or the PCCCH blocks. The MS performs a tradeoff between autonomy and downlink TBF establishment time when setting the NON_DRX_TIMER value.

Moreover, the MS is in non-DRX mode during the GPRS attach and RA update procedures because the BSS might not know the IMSI of this MS. Without knowing the IMSI, the BSS is not able to compute the MS PAGING_GROUP and thus is not able to forward paging toward the right paging subchannel. This case may occur when IMSI is unknown at the HLR level.

5.4.4 Paging Modes

The paging configuration may change in time according to partial traffic congestion in a given cell. Paging mode information is given by the network to the MSs in order to control possible additional requirements on paging decoding. Three paging modes are defined:

- *Normal paging.* The MS is required to decode all paging messages in the paging subchannel belonging to its paging group; this is the nominal case.

- *Extended paging* or "next but one." The MS is required to decode all paging messages in the paging subchannel belonging to its paging group and also in another paging subchannel. This mode is used when there is a temporary overload on some of the subchannels. If PBCCH and PCCCH are present in the cell, the MS decodes an additional subchannel that is located on the third PPCH after the PPCH corresponding to the MS paging group. If PBCCH and PCCCH are not present in the cell, the MS decodes an additional paging subchannel, which is identified as the next PCH after the paging subchannel corresponding to its paging group.

- *Paging reorganization.* The MS is required to decode all paging messages on (P)CCCH independently of its paging group. This mode is used by the network when it needs to change the logical channel organization between the (P)AGCH and (P)PCH.

5.4.5 Downlink Signaling Failure

A downlink signaling failure mechanism is implemented in GPRS mode as in GSM mode (see Figure 5.4). It is based on a *downlink signaling failure counter* (DSC), and it checks that the MS is able to successfully decode a message in its paging subchannel. The counter value is initialized with a DSC maximum value deduced from the paging period when the MS camps

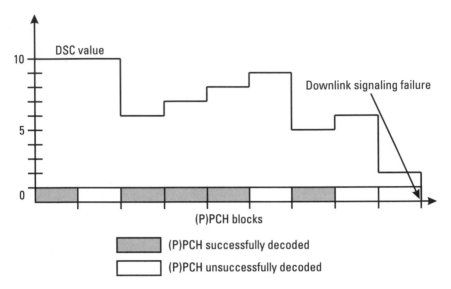

Figure 5.4   Downlink signaling failure mechanism.

on a new cell or leaves the packet transfer mode. The DSC value is incremented by one for each paging message successfully decoded, and decremented by four for each paging message unsuccessfully decoded. There is a downlink signaling failure if the DSC value is lower than zero. A cell reselection is triggered on downlink signaling failure.

5.5   Radio Resource Allocation

This section describes the procedures enabling the assignment of radio resource to the mobile and in consequence the establishment of a TBF. These procedures correspond to the transition from packet idle mode to packet transfer mode. The procedures used for the report of measurements are also detailed.

As described in Chapter 3, two kinds of TBF can be established:

- Uplink TBF assigning uplink radio resource to the mobile with a downlink signaling channel;
- Downlink TBF assigning downlink resources on shared PDCHs with an uplink signaling channel.

Two simultaneous TBFs assigned to the same MS in opposite direction (uplink and downlink) are said to be concurrent.

5.5.1 Uplink TBF Establishment

The mobile triggers the establishment of an uplink TBF for three major reasons:

- To perform an uplink data transfer;
- To answer a paging;
- To perform a GMM procedure (e.g., RA update procedure and GPRS attach procedure).

5.5.1.1 RACH/PRACH Phase

Access Request

The mobile triggers the establishment of an uplink TBF:

- By sending a CHANNEL REQUEST message on the RACH when there is no PBCCH in the cell;
- By sending a PACKET CHANNEL REQUEST message on the PRACH if PBCCH is present in the cell.

The mobile may also request the establishment of an uplink TBF during a downlink TBF on PACCH.

There are two codings for the PACKET CHANNEL REQUEST message. One format allows the transport of 8 bits information and the other one 11 bits information. The CHANNEL REQUEST message carries 8 bits of information.

The CHANNEL REQUEST message is used for both circuit-switched access and packet access. This message contains 8 bits of information, allowing a limited number of uplink combinations. As some of the combinations are reserved for GSM access, fewer combinations are available for packet access.

On PRACH, no circuit-switched access can occur. All the bit combinations of 8 or 11 bits of information within the access message are reserved for packet access only. Thus, access through the PRACH allows the mobile to send more precise information on its capability and requirements in order to establish as quickly as possible an uplink TBF.

Note On the RACH, the MS is unable to indicate its multislot capability in the CHANNEL REQUEST message. In order to use its maximum capability, it will be necessary to establish a TBF in two phases. The other possibility is to start the uplink transfer using only one time slot; once the mobile is identified, the BSS can then request the mobile capabilities to the SGSN and finally extend the allocation when they will be received. Even if the multislot capability of the mobile is given to the network during the GPRS attach procedure, the BSS does not have this information when it receives the access request. This information is only available at the SGSN side. The two-phase access procedure is described in the following sections.

As shown in Figure 5.5, the CHANNEL REQUEST and PACKET CHANNEL REQUEST messages are composed of two parts:

- One establishment cause;
- One random reference.

The random reference length is from 2 to 5 bits depending on the establishment cause.

The random reference reduces the probability that two MSs requiring the establishment of a TBF send exactly the same message in the same RACH or PRACH occurrence. This could happen, but the problem is solved by the contention resolution procedure described later.

When the MS requests the establishment of an uplink TBF, it randomly chooses the few bits that are sent within the random reference field of the access request message. When the network assigns the resources to the mobile it returns the request reference (random reference, establishment cause, and FN in which the AB was received) in order for the MS to correlate the assign-

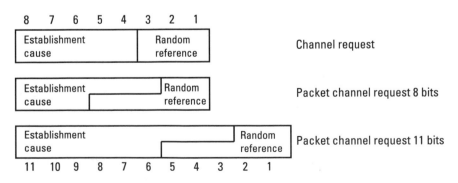

Figure 5.5 Channel request and packet channel request format.

ment and the request. If two mobiles have sent an AB at the same time using different request references, the one that is not addressed by the returned request reference value will abort the access procedure.

The two basic procedures used for the establishment of an uplink TBF are:

- The one-phase access procedure;
- The two-phase access procedure.

Which type of procedure is used is indicated or deduced from the establishment cause of the access request message.

The different establishment causes that can be signaled by the mobile within the PACKET CHANNEL REQUEST message are:

- One-phase access;
- Two-phase access;
- Short access (used when the mobile wants to transfer an amount of data that is less than 8 RLC/MAC blocks);
- Page response (used in response to a paging);
- Cell update (used to trigger a cell update procedure);
- MM procedure;
- Single block without TBF establishment (used to send a measure-ment report to the network).

For the short access, page response, cell update, and MM procedure access types, the TBF is established using one-phase access.

Within the CHANNEL REQUEST message, only two establishment causes are possible for packet access: the one-phase access and the single-block packet access. This last one is used to initiate the two-phase access procedure.

When the mobile requests a short access, one-phase access, or two-phase access using an 11-bits PACKET CHANNEL REQUEST message, it has to indicate the radio priority of the TBF. The highest radio priority is indicated when the TBF is used for signaling purposes.

If the mobile wants to establish a TBF in the RLC unacknowledged mode, it must request a two-phase access procedure. By default, the one-phase access requests a TBF in RLC acknowledged mode.

The two-phase access procedure is longer than the one-phase access, as it requires the exchange of four messages (see section below) rather than two as in the one-phase access.

The advantage of the two-phase access is that it allows the mobile to give more precision on its capabilities and requirements. As previously mentioned, during a one-phase access on the RACH, the mobile cannot provide its multislot capability. So in case of long uplink TBF, it is interesting for the mobile to request a two-phase access allowing a higher throughput thanks to the allocation of more time slots in uplink. On PRACH, the mobile can indicate its multislot class for a one-phase access within the PACKET CHANNEL REQUEST message.

Access Persistence Control on PRACH

It could happen that two MSs try to access the network at the same time and send their ABs on the same PRACH occurrence. In this situation, a collision is detected at the BTS side and if the two ABs are received with approximately the same power, neither of the two ABs is decoded. The access persistence control allows that the two mobiles avoid retransmitting ABs in the same PRACH occurrence in their next attempt. For that the occurrence in which the mobile will send the next AB is randomly determined. This mechanism is controlled dynamically by the network through the broadcast of access persistence control parameters on the PBCCH and PCCCH. The control of access persistence on PRACH is used to limit the collision probability on this channel.

The parameters involved in access persistence control are listed below.

- MAX_RETRANS indicates the maximum number of PACKET CHANNEL REQUEST message retransmissions the mobile is allowed to do.
- PERSISTENCE_LEVEL is a threshold whose usage is described below. The range of this parameter is {0, 1, 2, ... , 14, 16}.
- S indicates the minimum number of PRACH occurrences between two consecutive PACKET CHANNEL REQUEST messages.
- TX_INT defines the spreading interval (in terms of PRACH occurrences) of the random access.

Whenever the mobile attempts to send a PACKET CHANNEL REQUEST message, it has to draw a random value with uniform probability distribution in the set {0, 1, 2, ... , 15}. The MS is allowed to transmit a

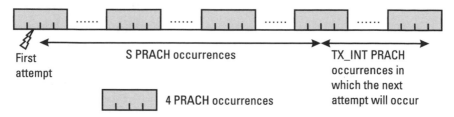

First attempt

S PRACH occurrences

TX_INT PRACH occurrences in which the next attempt will occur

4 PRACH occurrences

Figure 5.6 Access persistence control on PRACH.

PACKET CHANNEL REQUEST message if the random value is greater than or equal to PERSISTENCE_LEVEL. The next PRACH occurrence in which the MS attempts to send a PACKET CHANNEL REQUEST message is determined by the parameters S and TX_INT. This is illustrated in Figure 5.6. After MAX_RETRANS + 1 attempts to send a PACKET CHANNEL REQUEST message, the mobile is not allowed to transmit any more access messages.

Note that the PERSISTENCE_LEVEL and MAX_RETRANS parameters depend on the radio priority of the TBF. Four different radio priorities exist. Four values for each two parameters are broadcast by the network. The mobile chooses the one corresponding to the radio priority of the TBF it wants to establish.

Depending on the PRACH load and the number of collisions detected on the PRACH, the network can adjust the different access persistence control parameters, in order to regulate the load on PRACH and reduce the number of collisions. This ensures a constant throughput on the PRACH.

In order to reduce the load on the PRACH, the network can also forbid access to mobiles belonging to some access control classes. The access control class is a subscriber parameter. It is used to favor some subscribers in relation to others for accessing the network. The parameter ACC_CONTR_CLASS, which is broadcast on PBCCH, indicates which access control classes are allowed to access the network.

5.5.1.2 Uplink TBF Establishment on CCCH

When PBCCH is not present in the cell and a mobile in packet idle mode wants to establish an uplink TBF, it performs access on CCCH.

One-Phase Access Procedure on CCCH

Figure 5.7 describes the scenario for one-phase access uplink TBF establishment on CCCH. This procedure allows the allocation of only one time slot

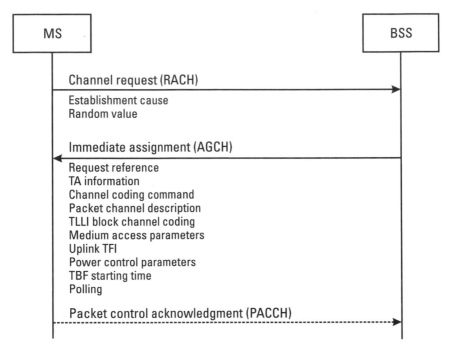

Figure 5.7   One-phase access establishment scenario on CCCH.

to the MS, even if its multislot class would have allowed more. This is due both to the impossibility of signaling the multislot class within the CHANNEL REQUEST message and to a limitation in the length of the IMMEDIATE ASSIGNMENT message.

The MS initiates this procedure by sending a CHANNEL REQUEST message on the RACH. The CHANNEL REQUEST message contains the establishment cause and the random value as described in the previous section. The establishment cause for this procedure is one-phase access.

Note that the network does not have to respect the one-phase access request of the mobile and may force a two-phase access procedure.

Upon reception of the CHANNEL REQUEST message, the BSS sends an IMMEDIATE ASSIGNMENT message on AGCH. This message contains the following information:

- *Request reference.* This includes the contents of the CHANNEL REQUEST message and the FN in which it was received.

- *TA parameters.* These include the TA index and the TA TN if continuous TA is implemented by the network and the initial TA. In this case the TA TN corresponds to the allocated PDCH.

- *Uplink TFI.* This parameter identifies the uplink TBF.

- *Channel coding command.* This parameter indicates to the mobile which coding scheme (CS-1, CS-2, CS-3, or CS-4) to use for uplink data transmission.

- *Packet channel description.* This indicates the allocated TN, the training sequence code, and the frequency parameters.

- *TLLI block channel coding.* This is used to indicate the coding scheme that must be used for data transmission during the contention resolution phase. The value could be CS-1 or the previous channel coding command. The contention resolution procedure is described later.

- *Power control parameters.* These indicate the downlink power control mode and the uplink power control parameters.

- *Medium access parameters.* These indicate the USF value in case of dynamic allocation and the fixed allocated bitmap in case of fixed allocation (refer to Section 3.3.2.3 for more details on the medium access methods).

Note that if the initial TA is not provided, the MS must await the reception of a correct TA value given during the continuous TA procedure before starting to transmit.

The assigned PDCH must be in the same frequency band as the BCCH, since the bands supported by the mobile are unknown at this time.

The network may request an acknowledgment from the mobile. The acknowledgment is requested by setting the polling bit in the IMMEDIATE ASSIGNMENT message. If the polling bit is set to 1, the mobile sends a PACKET CONTROL ACKNOWLEDGMENT message on the assigned PDCH, in the uplink block specified by the TBF starting time parameter. In this case the TBF starting time is used to indicate when the assigned PDCH becomes valid and when the uplink block for PACKET CONTROL ACKNOWLEDGMENT message is sent. The TBF starting time indicates the FN in which the uplink TBF starts.

The PACKET CONTROL ACKNOWLEDGMENT message is sent as either a normal RLC/MAC control block or as four consecutive identical

ABs. The format of this message depends on the CONTROL_ACK_TYPE parameter value broadcast on BCCH or PBCCH.

Note that the network may request the sending of a PACKET CONTROL ACKNOWLEDGMENT message in order to be sure that the mobile has received the uplink assignment and to avoid the allocation of uplink resources that will not be used, particularly in case of dynamic allocation.

At the end of the assignment procedure, the mobile enters into contention resolution phase. This procedure is described later in this chapter.

Two-Phase Access Procedure on CCCH

The mobile must initiate a two-phase access procedure when it wants to establish a TBF in RLC unacknowledged mode. This procedure may be requested in other cases not specified in the GSM standard (e.g., if the mobile wants to establish an uplink TBF with multislot allocated on CCCH).

Note that when a two-phase access is requested by the mobile, the network must use this procedure to establish the uplink TBF.

Figure 5.8 describes the scenario for two-phase access uplink establishment on CCCH.

The mobile initiates the two-phase access procedure by sending a CHANNEL REQUEST message on RACH requesting a single-block packet access. Upon receipt of the CHANNEL REQUEST message, the network sends an IMMEDIATE ASSIGNMENT message to the mobile on AGCH.

This message allocates one single uplink block to the mobile on a PDCH.

The IMMEDIATE ASSIGNMENT message contains the following parameters:

- *Request reference.* This includes the contents of the CHANNEL REQUEST message and the FN in which it was received.
- *Packet channel description.* This indicates the allocated TN, the training sequence code, and the frequency parameters.
- *Power control parameters.* These indicate the downlink power control mode and the uplink power control parameters.
- *TBF starting time.* This indicates the FN in which the mobile will start sending its uplink single block.
- *Initial TA.* This is used to transmit in the single block allocation.

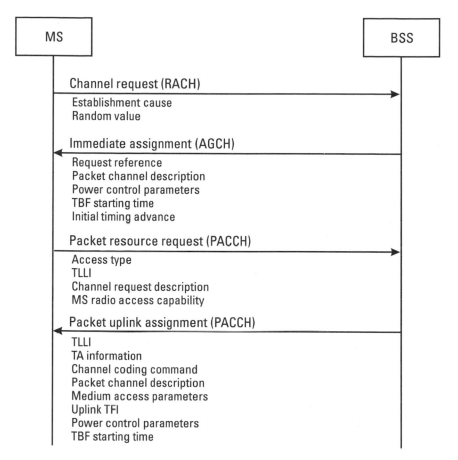

Figure 5.8 Two-phase access establishment scenario on CCCH.

Up to this point, no TFI is allocated to the mobile. The network does not know the exact reason for the establishment cause (the mobile may have requested a single-block packet access for the mere purpose of sending a measurement report to the network; see Section 5.5.3).

In this uplink block occurrence, the mobile sends a PACKET RESOURCE REQUEST message in order to indicate the two-phase access request. It contains the following information:

- *TLLI.* This parameter uniquely identifies the mobile.
- *Access type.* This indicates the reason for requesting the access (in this example, two-phase access).

- *Channel request description.* This indicates the peak throughput class for the PDP context of the LLC PDU, the radio priority, the RLC mode, the type of the first LLC PDU, and the number of RLC data octets of the requested TBF.

- *MS radio access capability.* This indicates the mobile capabilities in terms of multislot class and RF power for the different frequency bands that are supported by the mobile.

With all this information, the network is able to allocate uplink resources in a more efficient way. The requested TBF properties are used by the network to select the multiplexing level matching. Depending on the QoS parameters, more or less bandwidth will be allocated to the mobile. The mobile that will share the same uplink PDCHs as other mobiles will have more or less uplink resources depending on these parameters. The radio access capability parameters are used to take maximum advantage of the mobile capabilities. The allocated PDCHs can be in any frequency band that is supported by the mobile.

Note that if the mobile indicates 0 for the number of RLC data octets that have to be transferred, the network interprets the TBF as an open-ended TBF. In this case, the number of uplink resources needed is undetermined. This information concerns only the fixed-allocation multiplexing scheme for which it is very important for the network to know the exact amount of uplink data in order to optimize the allocation of uplink resources. At this time in this procedure, the mobile does not know the uplink multiplexing scheme that is used by the network. Thus this information will always be provided.

Upon receipt of the PACKET RESOURCE REQUEST message, the BSS sends a PACKET UPLINK ASSIGNMENT message on PACCH assigning uplink resources to the mobile. This message contains the following parameters:

- *TLLI.* This is used to address the block in downlink.

- *TA parameters.* These include the TA index and the TA TN if continuous TA is implemented by the network.

- *Uplink TFI.* This parameter identifies the uplink TBF.

- *Channel coding command.* This parameter indicates to the mobile which coding scheme (CS-1, CS-2, CS-3, or CS-4) to use for uplink data transmission.

- *Packet channel description*. This indicates the training sequence code and the frequency parameters.

- *Power control parameters*. These indicate the downlink power control mode and the uplink power control parameters.

- *Medium access parameters*. These include USF values on the different allocated time slots in case of dynamic allocation or extended dynamic allocation, and time slot allocation, downlink control timeslot (downlink PACCH), and the fixed-allocation bitmap parameters in case of fixed allocation.

The network provides a TBF starting time in case of fixed allocation. It indicates the FN in which the bitmap starts to be valid. In case of dynamic or extended dynamic allocation, the TBF starting time is optional. If it is provided, the mobile starts listening to the USF in the FN indicated.

The network can poll the mobile within the PACKET UPLINK ASSIGNMENT message to request the sending of a PACKET CONTROL ACKNOWLEDGMENT message. The polling is performed using the S/P and RRBP fields of the downlink control block MAC header.

5.5.1.3   Uplink TBF Establishment on PCCCH

When PBCCH is present in the cell and a mobile in packet idle mode wants to establish an uplink TBF, it performs access on PCCCH.

One-Phase Access Procedure on PCCCH

Figure 5.9 describes the scenario for one-phase access uplink establishment on PCCCH. The mobile triggers the establishment of an uplink TBF using the one-phase access procedure by sending a PACKET CHANNEL REQUEST message on PRACH indicating one of the following access types: one-phase access, short access, page response, cell update, or MM procedure.

Note that despite the request of the mobile, the network can force a two-phase access procedure when a one-phase access is requested.

The PACKET CHANNEL REQUEST message contains either 8 or 11 bits of information depending on the ACCESS_BURST_TYPE parameter, which is broadcast on PBCCH. The establishment cause of the PACKET CHANNEL REQUEST message contains the multislot class of the mobile. Moreover, if the message is sent in the 11-bit format, it also contains the radio priority of the TBF. The network is then able to assign uplink resources matching the multislot capability of the mobile.

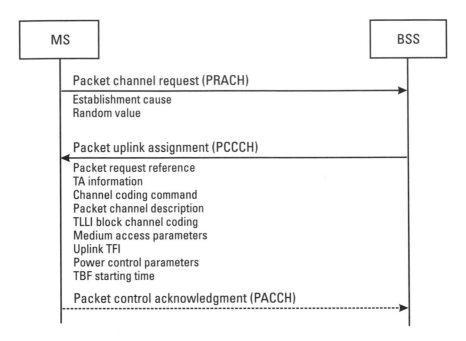

Figure 5.9   One-phase access establishment scenario on PCCCH.

The radio priority parameter, if available, can be used by the network to prioritize the request in relation to others received from other mobiles. The network sends the PACKET UPLINK ASSIGNMENT message on the PCCCH on which the request has been received. This message contains the same parameters as described in Section 5.5.1.2 for one-phase access procedure on CCCH, except that the packet request reference is given instead of the request reference and the medium access parameters handle more than one time slot allocation.

The medium access parameters consists of:

- USF values on the different allocated time slots in case of dynamic allocation or extended dynamic allocation;
- Time slot allocation, downlink control time slot, and the fixed-allocation bitmap parameters in case of fixed allocation.

The allocated PDCH will be in the same frequency band as the PBCCH because the radio access technology types of the mobile are not known at this time.

When the mobile receives the PACKET UPLINK ASSIGNMENT message, it enters into packet transfer mode and the contention resolution phase starts. The network can poll the mobile within the PACKET UPLINK ASSIGNMENT message to request the sending of a PACKET CONTROL ACKNOWLEDGMENT message. The polling is performed by using the S/P and RRBP fields of the downlink control block MAC header.

Two-Phase Access Procedure on PCCCH

The mobile requests a two-phase access request when it wants to establish a TBF in RLC unacknowledged mode. This procedure may also be requested by the mobile to establish a TBF in RLC acknowledged mode. Figure 5.10

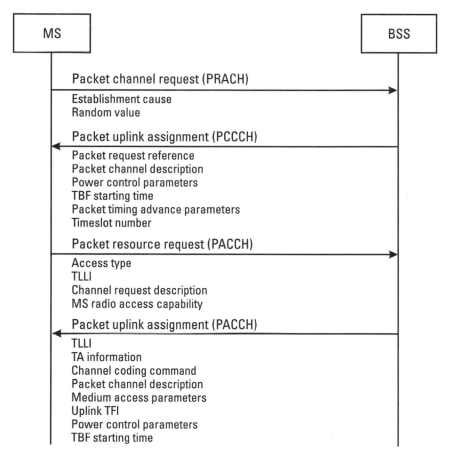

Figure 5.10 Two-phase access establishment scenario on PCCCH.

describes the scenario for two-phase access uplink TBF establishment on PCCCH when the mobile is in packet idle mode.

The MS requests the establishment of a TBF in two-phase access by sending a PACKET CHANNEL REQUEST message on PRACH. The establishment cause within the PACKET CHANNEL REQUEST message indicates two-phase access.

Note that the network must respect the request of the mobile.

If the AB is sent with the 11-bit format, the radio priority is also indicated within the establishment cause.

On receipt of the access request, the network allocates a single uplink block to the mobile by sending a PACKET UPLINK ASSIGNMENT message. This message contains the following information:

- *Packet request reference*. This includes the contents of the PACKET CHANNEL REQUEST message and the FN in which it was received.

- *Packet channel description*. This indicates the training sequence code and the frequency parameters.

- *Power control parameters*. These indicate the downlink power control mode and the uplink power control parameters.

- *TBF starting time*. This indicates the FN in which the mobile will send the PACKET RESOURCE REQUEST message.

- *Packet TA parameters*. The initial TA is given to transmit in the single block allocation.

- *Time slot number*. This indicates the time slot on which the block must be sent.

The end of the procedure is exactly the same as the one for two-phase establishment on CCCH (refer to Section 5.5.1.2).

Packet Access Queuing Notification Procedure

Whenever the network receives an access request from a mobile on PRACH and it cannot satisfy the request, a queuing notification may be sent to the mobile. This will indicate to the mobile that its request has been correctly received by the network and taken into account, but the network is temporarily unable to satisfy it. From a network point of view, this procedure reduces PRACH congestion. When the mobile receives the queuing notification, it will stop trying to access the cell. This procedure is only supported on PCCCH.

Figure 5.11 describes an example of uplink access establishment preceded by a temporary queuing phase.

The mobile requests the establishment of a TBF by sending a PACKET CHANNEL REQUEST message on PRACH. Because of congestion or other reasons, the network is unable to satisfy the request. A PACKET QUEUING NOTIFICATION message is sent on the PCCCH on which the PACKET CHANNEL REQUEST message was received.

The PACKET QUEUING NOTIFICATION message contains the following information:

- *Packet request reference.* This includes the contents of the PACKET CHANNEL REQUEST message and the FN in which it was received.

- *Temporary queuing identifier (TQI).* This parameter identifies the request of the mobile.

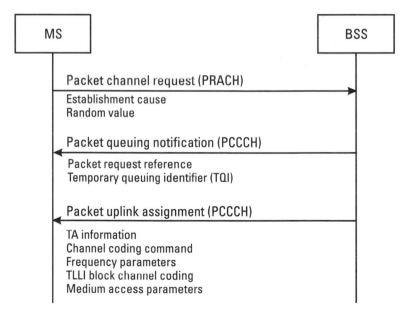

Figure 5.11 Packet queuing notification procedure.

When the mobile receives the queuing indication, it starts monitoring its downlink PCCCH. The queuing duration is handled by a timer at the MS and BSS side.

Whenever uplink resources again become available at network side, the BSS sends a PACKET UPLINK ASSIGNMENT message to the mobile assigning uplink resource corresponding to the mobile requests. The mobile is addressed within the PACKET UPLINK ASSIGNMENT message using the TQI.

5.5.1.4   Uplink TBF Establishment When Downlink TBF Is Already Established

The mobile has the possibility of establishing an uplink TBF when it is in packet transfer mode, thus during a downlink TBF. Figure 5.12 describes this procedure.

The mobile requests the establishment of an uplink transfer during a downlink TBF by including a channel request description within the PACKET DOWNLINK ACK/NACK message. This message is used by the mobile to acknowledge the RLC data blocks received in downlink. Its sending on the PACCH is controlled by the polling mechanism on a network order.

The channel request description contains the same information as described in Section 5.5.1.2 for the two-phase access procedure on CCCH.

When the network detects the channel request description within a PACKET DOWNLINK ACK/NACK message, it allocates uplink resources

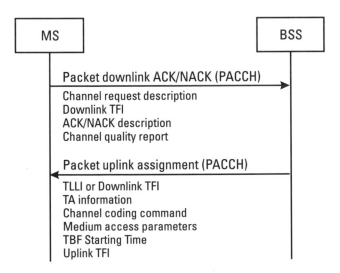

Figure 5.12 Procedure for uplink establishment when the MS is in Packet Transfer Mode.

(if available) to the mobile by sending a PACKET UPLINK ASSIGN-MENT message, which contains the following information:

- *Downlink TFI or TLLI.* This parameter is used to address the mobile in downlink.

- *TA information.* In case of continuous TA, the network may allocate a new time slot and TAI for the procedure.

- *Channel coding command.* This indicates the coding scheme for use in the uplink direction.

- *Medium access parameters.* This was described in Section 5.5.1.2 for the two-phase access procedure on CCCH.

- *TBF starting time.* This indicates the FN in which the uplink TBF starts.

- *Uplink TFI.* This identifies the uplink TBF.

Note that:

1. When the BSS allocates the uplink resources it must respect the multislot class of the mobile. As a downlink TBF already exists, the uplink resources assigned must be compatible with the downlink ones. If the network is not able to allocate uplink resources matching the downlink ones, it can reallocate at the same time the uplink and downlink resources by using the PACKET TIMESLOT RECONFIGURE message. This message allows reallocation of both uplink and downlink resources.

2. When the MS receives the PACKET UPLINK ASSIGNMENT or PACKET TIMESLOT RECONFIGURE message, it will not enter in contention resolution phase since the mobile is already uniquely identified at the BSS side.

3. The network could reject the request by sending a PACKET ACCESS REJECT message on the PACCH.

5.5.1.5   Uplink TBF Modification

The network can modify at any time the uplink TBF by sending either a new PACKET UPLINK ASSIGNMENT message as illustrated in Figure 5.13 or a PACKET TIMESLOT RECONFIGURE message on the PACCH. The PACKET TIMESLOT RECONFIGURE message is only used when there

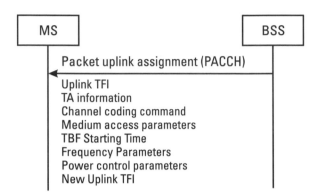

Figure 5.13 Modification of the uplink TBF initiated by the network.

are two concurrent TBFs already established. Through this procedure, the network can modify the multislot allocation of the mobile.

Such a procedure could be triggered when there is a change of service requested by the MS. In fact, during an uplink transfer, upper layers may request the transfer of an LLC PDU with a different radio priority or peak throughput class. The modification of the service demand is requested by the mobile by sending a PACKET RESOURCE REQUEST message on the PACCH including the channel request description.

This procedure can also be triggered by the network in the following situations:

- In case of modification of the allocated PDCHs in the cell (e.g., increase of circuit-switched traffic and thus reduction of the number of PDCHs in the cell);
- In case the link quality on the allocated PDCHs for the mobile is not sufficient.

The network can provide a new TFI assignment. The new allocation takes effect in the FN indicated by the TBF starting time.

5.5.1.6   Contention Resolution Procedure

It could happen that two mobiles trying to access the network send the same CHANNEL REQUEST message (respectively PACKET CHANNEL REQUEST message) within the same RACH occurrence (respectively PRACH occurrence). It may also happen that the two mobiles use the same establishment cause and the same packet reference request.

If the two messages are received at the BTS side with approximately the same level, the BTS will detect a collision on RACH or PRACH and then no access message will be decoded. The two mobiles will try to access the network later on, and the probability that the same occurrence is chosen is low. This does not cause any problem.

However, if one mobile is near the BTS and the other one is far away, due to the capture effect there is a high probability that the BTS decodes the message that was sent by the nearest mobile. In this case, the BTS will allocate uplink resources by sending a PACKET UPLINK ASSIGNMENT message. Since the two mobiles have transmitted the same message in the same RACH or PRACH occurrence, they will both identify their packet reference request and behave as if the assignment message was for them. In this case the two MSs will transmit in the same uplink PDTCH occurrence as described in Figure 5.14.

In order to avoid having one of the two MSs receive data that is not intended for it, the mobile, during the contention resolution phase, must not accept the establishment of a concurrent TBF. It must also not accept control messages such as PACKET CELL CHANGE ORDER message and PACKET POWER CONTROL/TIMING ADVANCE. For security reason, this shall be avoided.

Contention Resolution at One-Phase Access

In order to avoid this problem after uplink TBF establishment, the MS must insert its TLLI within each uplink RLC data blocks until the contention res-

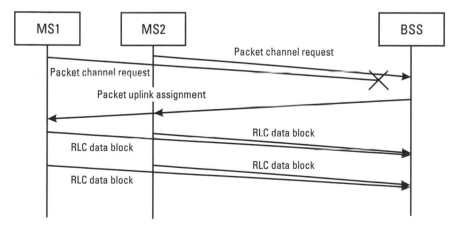

Figure 5.14 Contention at TBF establishment.

olution period is finished (see Figure 5.15). The TLLI uniquely identifies an MS within one RA.

The uplink RLC data blocks, which contain the TLLI, are coded using the TLLI channel coding command, which is indicated in the assignment message. At the decoding of the first uplink RLC data block including a TLLI, the network sends a PACKET UPLINK ACK/NACK with the TLLI that was received in the uplink RLC data block. At this time, the contention resolution is completed at network side.

If two mobiles had performed an access on the same RACH or PRACH occurrence and sent the same message, the contention resolution would fail for the mobile that received the PACKET UPLINK ACK/NACK message including its assigned TFI but with a TLLI value other than that which the MS included in the RLC header of the uplink data blocks.

The contention resolution is completed at the mobile side when the mobile to which the TLLI belongs receives the PACKET UPLINK ACK/ NACK message. It continues the uplink transfer without including the TLLI in the uplink RLC data blocks and using the channel coding command for data block encoding.

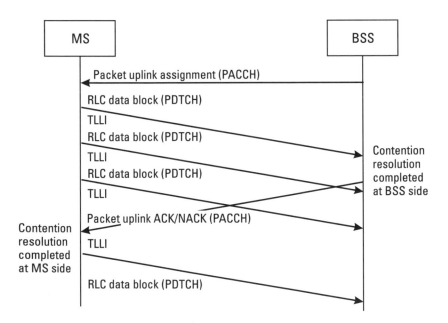

Figure 5.15 Contention resolution at one-phase access.

Contention Resolution at Two-Phase Access

In this case, the contention resolution is completed on the network side when it receives the PACKET RESOURCE REQUEST message including the TLLI. The contention resolution is completed at mobile side when it receives the second uplink assignment message with its TLLI included. The contention resolution fails at MS side when it receives the second assignment message with a TLLI value other than that which the MS has included in the PACKET RESOURCE REQUEST message.

5.5.2 Downlink TBF Establishment

The establishment of a downlink TBF is performed on the CCCH if there is no PBCCH in the cell; otherwise, it is performed on PCCCH. It can also be performed on PACCH during an uplink transfer.

The BSS initiates the establishment of a downlink TBF when it receives a downlink LLC PDU from the SGSN that must be transmitted to a mobile, and this MS is not already in downlink transfer. The transmission of an LLC PDU from the SGSN to the BSS is only allowed when the mobile is located at cell level within the SGSN. If this is not the case, the SGSN must start a paging procedure from which it will recover this information (see Chapter 3). At the end of the paging procedure, the mobile is in GMM READY state.

When the SGSN transmits a downlink LLC PDU that must be sent to the mobile, it indicates to the BSS the cell in which the mobile is located, the TLLI identifying the mobile, the MS radio access capability parameters (indicating the multislot class of the MS, the RF power capability, and the supported frequency bands) and the QoS profile. These parameters are used by the BSS to address the mobile and derive the downlink allocation for the transfer of the LLC PDU.

5.5.2.1 Downlink TBF Establishment on CCCH

Figure 5.16 describes an example of downlink TBF establishment on CCCH when the mobile is in packet idle mode.

When the network receives a downlink LLC PDU to transmit to the mobile, it initiates the establishment of a downlink TBF by sending an IMMEDIATE ASSIGNMENT message to the MS on CCCH. The IMMEDIATE ASSIGNMENT message is sent on any block of the CCCH if the mobile is in non-DRX mode; otherwise, it is sent on one block corresponding to the paging group of the mobile.

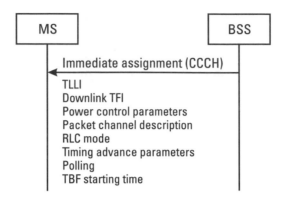

Figure 5.16 Downlink TBF establishment on CCCH.

The IMMEDIATE ASSIGNMENT message contains the following parameters:

- *TLLI.* This indicates the identity of the mobile for which the message is intended.
- *Downlink TFI.* This is the identifier of the downlink TBF.
- *Power control parameters.* These indicate the downlink power control mode and the uplink power control parameters.
- *Packet channel description.* This indicates the allocated TN, the TS code, and the frequency parameters.
- *RLC mode.* This indicates whether the TBF is in RLC acknowledged mode or RLC unacknowledged mode.
- *TA parameters.* These include the TA index and the TA TN if continuous TA is implemented by the network. The initial TA may not be present, as the network has no information on it except perhaps from a previous TBF.

Within the IMMEDIATE ASSIGNMENT message, the BSS is not able to allocate more than one time slot in downlink to the MS despite the fact that it knows the MS multislot class received from the SGSN. This is due to a limitation in the IMMEDIATE ASSIGNMENT message size.

Initial TA Problem

In most cases, no initial TA value can be provided to the mobile in the first assignment message. The mobile is not allowed to transmit before having received a correct TA value. If the continuous TA procedure is used, it may take up to two seconds before the MS receives a usable TA value and then is allowed to transfer in uplink. This could delay the TBF.

In order to accelerate the acquisition of the initial TA value, the network can request the sending of a PACKET CONTROL ACKNOWL-EDGMENT message by setting the polling bit to 1 in the IMMEDIATE ASSIGNMENT message.

The mobile sends the PACKET CONTROL ACKNOWLEDG-MENT message on the assigned PDCH, in the uplink block specified by the TBF starting time parameter. The TBF starting time indicates when the assigned PDCH becomes valid for the downlink transfer.

The PACKET CONTROL ACKNOWLEDGMENT message can be sent as either a normal RLC/MAC control block or as four consecutive identical ABs. The format of this message depends on the CONTROL_ACK_TYPE parameter value, broadcast on BCCH or PBCCH.

If the CONTROL_ACK_TYPE indicates AB type, the mobile sends the four consecutive ABs with a TA of zero. The BTS deduces the initial TA at the reception of these ABs. This estimation can be sent in a PACKET POWER CONTROL/TIMING ADVANCE message. The network will not use the combination polling plus CONTROL_ACK_TYPE set to AB type, as the mobile is not allowed to transmit without a valid TA value. Figure 5.17 describes this procedure.

Note that another solution consists of sending a PACKET DOWN-LINK ASSIGNMENT message rather than sending a PACKET POWER CONTROL/TIMING ADVANCE message. This message can be used to provide both the initial TA and a PDCH allocation that takes advantage of the MS multislot class.

5.5.2.2 Downlink TBF Establishment on PCCCH

Figure 5.18 describes an example of downlink TBF establishment on PCCCH when the mobile is in packet idle mode.

When the network receives a downlink LLC PDU to transmit to the mobile, it initiates the establishment of a downlink TBF by sending a PACKET DOWNLINK ASSIGNMENT message to the MS on PCCCH. The PACKET DOWNLINK ASSIGNMENT message is sent on any block of the PCCCH where paging may appear if the mobile is in non-DRX mode; otherwise, it is sent on one block corresponding to the paging group of the mobile.

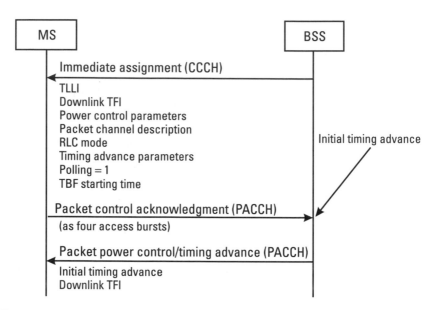

Figure 5.17 Downlink TBF establishment with initial TA computation.

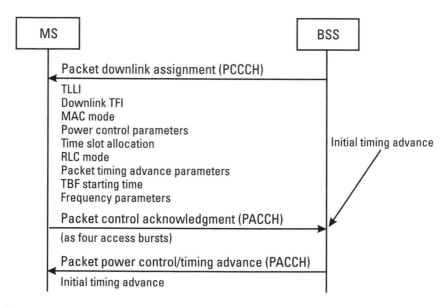

Figure 5.18 Example of downlink TBF establishment on PCCCH.

The PACKET DOWNLINK ASSIGNMENT message contains the following parameters:

- *TLLI.* This identifies the mobile for which the message is intended.
- *Downlink TFI.* This is the identifier of the downlink TBF.
- *Power control parameters.* These indicate the downlink power control mode and the uplink power control parameters.
- *MAC mode.* This indicates the medium access method, which will be used if an uplink TBF is established during this downlink transfer.
- *RLC mode.* This indicates whether the TBF is in RLC acknowledged mode or RLC unacknowledged mode.
- *Time slot allocation.* This indicates the time slot assigned to the downlink allocation.
- *Packet TA parameters:* These include the TA index and the TA TN if continuous TA is implemented by the network. The initial TA may not be present because the network has no information on it except perhaps from a previous TBF.
- *Frequency parameters.* These give the frequency or the list of frequencies that are used on PDCHs.
- *TBF starting time.* If present, this indicates the FN in which the downlink TBF starts.

The BSS polls the mobile within the PACKET DOWNLINK ASSIGNMENT message requesting the sending of a PACKET CONTROL ACKNOWLEDGMENT message in four ABs. The polling is performed by using the S/P and RRBP fields of the downlink control block MAC header.

The BTS computes the initial TA that is indicated to the mobile by sending a PACKET POWER CONTROL/TIMING ADVANCE message on PACCH.

5.5.2.3 Downlink TBF Establishment When Uplink TBF Is Already Established

Whenever the BSS receives a downlink LLC PDU to transmit to a mobile for which an uplink TBF is already established, the BSS initiates the downlink TBF on the PACCH of the uplink TBF.

The BSS can initiate the downlink TBF establishment by sending either a PACKET DOWNLINK ASSIGNMENT message or a PACKET TIMESLOT RECONFIGURE message. The first allocates downlink

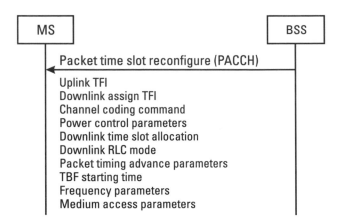

Figure 5.19 Packet downlink establishment on the PACCH.

resources without changing the uplink allocation. The second allows the network to change the uplink allocation if necessary and allocate downlink resources.

Figure 5.19 describes an example of downlink TBF establishment when the mobile is in packet transfer mode.

The BSS sends a PACKET TIMESLOT RECONFIGURE message on the downlink PACCH in order to establish a downlink TBF. This message can be used to reassign uplink resources to the mobile.

The PACKET TIMESLOT RECONFIGURE message contains the following information:

- *Uplink TFI.* This is used to identify the uplink TBF; in this case, to address the mobile on the downlink PACCH.

- *Downlink assign TFI.* This is the assigned TFI for the downlink TBF.

- *Channel coding command.* This parameter indicates to the mobile which coding scheme (CS-1, CS-2, CS-3, or CS-4) is to be used for uplink data transmission.

- *Power control parameters.* These indicate the downlink power control mode and the uplink power control parameters.

- *Downlink time slot allocation.* This is the PDCH allocation for the downlink transfer.

- *Downlink RLC mode.* This indicates either RLC acknowledged mode or RLC unacknowledged mode.

- *Packet TA parameters*. Includes the TA index and the TA TN if continuous TA is implemented by the network. and optionally the initial TA.

- *TBF starting time*. This indicates in which FN the resource reassignment takes place.

- *Frequency parameters*. These parameters are optionally given if the network wants to change the frequency configuration of the PDCHs.

- *Medium access parameters*. These describe the uplink access method, either fixed allocation or dynamic allocation or extended dynamic allocation.

5.5.2.4   Downlink TBF Modification

The BSS initiates resource reassignment by sending a PACKET DOWN-LINK ASSIGNMENT message or a PACKET TIMESLOT RECONFIG-URE message on PACCH. The PACKET TIMESLOT RECONFIGURE message is only sent when there is both an uplink and downlink TBF established (or the network wants to establish an uplink TBF during downlink transfer).

During this procedure, the network is not allowed to change the RLC mode or the TBF mode (medium access method). A change of RLC mode or TBF mode can only occur after a release of the TBF.

The resource reassignment takes effect in the FN indicated by the TBF starting time of the assignment message.

5.5.3   Measurement Report in Packet Idle Mode

In packet idle mode, as long as the mobile is in GMM READY state and the cell reselection mode is NC1 or NC2, the mobile regularly reports measurements to the network. The MEASUREMENT REPORT message fits in one radio block. As the importance of this message is not vital for the system (single messages can be lost), it has been decided to transfer this message without TBF establishment and without acknowledgment of the message (no RLC procedure is used). In order to transfer its measurements, the mobile simply requests one uplink radio block occurrence from the network.

5.5.3.1   Measurement Report Procedure on CCCH

When the MS is in packet idle mode and it must send a PACKET MEA-SUREMENT REPORT message, it requests via the CHANNEL

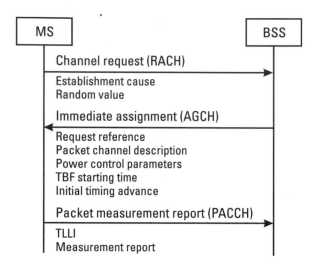

Figure 5.20 Measurement report procedure on CCCH in packet idle mode.

REQUEST message a single block packet access. Upon receipt of the request, the network sends an IMMEDIATE ASSIGNMENT message, assigning one single uplink block to the mobile. The mobile sends a PACKET MEASUREMENT REPORT message in this occurrence. The information that is provided to the mobile in the assignment message is the

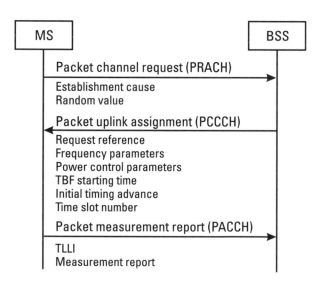

Figure 5.21 Measurement report procedure on PCCCH in packet idle mode.

same as that in the first assignment message during the two-phase access on CCCH. Figure 5.20 describes the measurement report procedure on CCCH.

5.5.3.2 Measurement Report Procedure on PCCCH

On PCCCH, the mobile initiates the measurement report procedure by sending a PACKET CHANNEL REQUEST message indicating "single block without TBF establishment" as the establishment cause.

The BSS allocates one uplink block occurrence by sending a PACKET UPLINK ASSIGNMENT message. The mobile transmits a PACKET MEASUREMENT REPORT message in the allocated uplink occurrence. Figure 5.21 describes this procedure.

5.6 RLC

The RLC layer provides a reliable link between the BSS and the MS, allowing transmission of RLC blocks in acknowledged or unacknowledged mode during an uplink or downlink TBF. The RLC layer is responsible for the segmentation of LLC frames into RLC data blocks. Before being transmitted on the radio interface, these blocks are numbered by the transmitter so that the receiver is able to detect undecoded data blocks and request their selective retransmission. When all the RLC data blocks belonging to one LLC frame have been received, the RLC layer at the receiver side ensures the reassembly of the LLC frame.

The RLC layer also provides a similar mechanism for the transmission of RLC/MAC control messages from the BSS to the mobile.

5.6.1 Transmission Modes

The RLC *automatic repeat request* (ARQ) functions support two modes of operation:

- RLC acknowledged mode;
- RLC unacknowledged mode.

In RLC acknowledged mode, the RLC ensures the selective retransmission of RLC data blocks that have not been correctly decoded by the receiver. This mode is used to achieve a high reliability in LLC PDU sending. In RLC unacknowledged mode, RLC data blocks that have not been correctly decoded are not retransmitted by the sending entity. This mode is used for applications that are tolerant of error and that request a constant throughput, such as streaming applications (video or audio streaming).

In uplink, the requested RLC mode is indicated in the PACKET RESOURCE REQUEST message in case of two-phase access and is by default RLC acknowledged mode in case of one-phase access. In downlink, the RLC mode is indicated within the assignment message by the RLC_MODE parameter. During the TBF, the RLC mode cannot be changed. Any change in the RLC mode previously requires the release of the TBF.

5.6.2   Segmentation and Reassembly of LLC PDUs

5.6.2.1   Segmentation of LLC PDUs into RLC Data Blocks

The segmentation of LLC PDUs allows the transport through the radio interface of LLC PDUs, whose size can be much larger than the data unit length of a single RLC data block. The segmentation of LLC PDUs is done in a dynamic way. It means that a change of coding scheme bringing about the decrease or the increase of the RLC data block unit length could happen at any time during the transmission. Depending on the CS used to encode the transmit RLC data block, the LLC PDU will be segmented in variable size data units as described in Figure 5.22.

The LLC PDUs are segmented in the same order as they are received by the transmit entity. Each RLC data block resulting from the segmentation of an LLC PDU is numbered using the BSN field of the RLC header. The BSN ranges from 0 to 127.

For one TBF, the first segmented RLC data block of the first LLC frame is numbered with BSN = 0, the next one with BSN = 1, and so on. The RLC data blocks are numbered modulo 128.

As described in Figure 5.22, if the contents of an LLC PDU do not fill an integer number of RLC data blocks, the beginning of the next LLC PDU and the end of the previous LLC PDU are placed in the same radio block. The LI, extension (E), and more (M) fields of the RLC data block header are used to delimit the consecutive LLC PDUs within one RLC data block.

The E field is used to indicate the presence of an optional extension octet in the RLC data block header. When set to 1 the E bit indicates that no extension octet follows. The M field is used to indicate (when set to 1) the presence of a new LLC PDU that starts after the current LLC PDU.

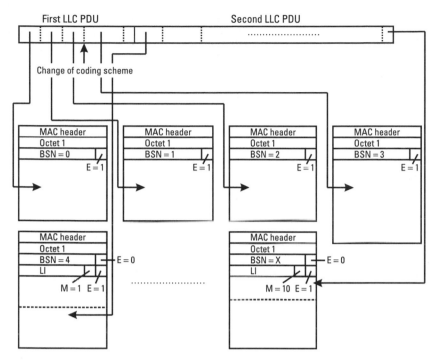

Figure 5.22 Segmentation mechanism.

The combination M = 0 and E = 1 indicates that there is no LLC PDU after the current one and there are no more extension octets. The combination M = 1 and E = 0 indicates that a new LLC PDU starts after the current one and there is an extension octet that delimits the new one. The combination M = 1 and E = 1 indicates that a new LLC PDU starts after the current one and continues until the end of the block. The LI field is used to delimit the LLC PDUs within the RLC data block. The first LI of the RLC data block indicates the number of octets of the RLC data unit belonging to the first LLC PDU.

Figure 5.22 provides an example of how the E, M, and LI fields are used for LLC PDUs delimitation within RLC data blocks.

5.6.2.2 Reassembly of LLC PDUs from RLC Data Blocks

After having received all the RLC data blocks belonging to an LLC PDU, the reassembly involves removing the RLC header of each RLC data block and reassembling the data unit using the BSN sequencing.

In case of RLC unacknowledged mode, some RLC data blocks may not have been decoded during the transfer. The RLC data units not received must be substituted with fill bits having the value 0.

5.6.3   Transfer of RLC Data Blocks

5.6.3.1   In Acknowledged RLC Mode

Sliding Window Mechanism

The transfer of RLC data blocks in the acknowledged RLC mode is controlled by a selective ARQ mechanism. This retransmission mechanism relies on the identification of each RLC data block, thanks to its BSN.

The RLC protocol relies on a sliding window mechanism. The size of this window for GPRS is 64 blocks. This window ensures that the gap, in term of block number, between the oldest unacknowledged block and the next-in-sequence block to be transmitted is always lower than 64. The sending side is not allowed to transmit a block when its BSN modulo 128 is higher than (BSN of the oldest unacknowledged block + 64) modulo 128. In this case, the RLC protocol is stalled. In this situation, the transmit entity is only allowed to retransmit the RLC data blocks that have not yet been acknowledged.

At the beginning of the TBF, the transmitter starts sending RLC data blocks in sequence. The oldest unacknowledged block is then the one numbered with BSN = 0. New RLC data blocks can be sent as long as their BSN is lower than 63.

Each time an acknowledgment message is received from the receiver, the transmitter updates its list of unacknowledged RLC blocks and starts their retransmission from the oldest unacknowledged one. When the acknowledgement message indicates that the oldest unacknowledged block has been received and decoded at receiver side, the RLC window slides up to the next-oldest unacknowledged block indicated by the message. If the protocol was previously stalled, it is not anymore.

The receiver maintains a table that contains the state of all the RLC data blocks within the window. At the beginning of the TBF, the receiver expects to receive the block with BSN = 0. Since the RLC blocks are sent in sequence by the sending side, the receiver can deduce from a correctly decoded block all the previously sent blocks that have not been decoded. Whenever an RLC block is decoded, it is marked as "decoded" within the table.

In case of a downlink TBF (the sending entity is the network), the PACKET DOWNLINK ACK/NACK message that contains the information on which blocks have not been decoded at the mobile side is always sent by the mobile on a BSS order (polling). In case of uplink TBF, the decision to send a PACKET UPLINK ACK/NACK message is also made by the network.

Each such message acknowledges all correctly received RLC data blocks up to an indicated BSN, thus "moving" the beginning of the sending window on the transmit side. Additionally, a bitmap is used to indicate BSNs of erroneously received RLC data blocks. The sending side then retransmits the erroneous RLC data blocks.

Figure 5.23 shows how the sliding window mechanism works on the transmit side as well as on the receive side. In the first state of the transmitter, the RLC data blocks with BSN 120, 12, and 37 have been transmitted but are unacknowledged. The next block that will be transmitted has BSN = 38. The receiver has not decoded the block with BSN = 12 and the last correctly decoded block was BSN = 36. The RLC data block with BSN = 37 that has been previously sent is not decoded by the receiver.

The transmitter sends the RLC block with BSN = 38 that is decoded by the receiver. This latter detects that the RLC block with BSN = 37 has not been decoded and BSN = 38 is the newest received RLC block.

At the next step, the receiver sends an acknowledgment message indicating that all the blocks until BSN = 38 have been received except BSN = 12 and BSN = 37. Upon receipt of this message, the transmitter updates its state table and the window is moved up to BSN = 12. The transmitter retransmits the block with BSN = 12. This block is decoded by the receiver that updates its received state table.

RLC Data Block Transfer During Uplink TBF

The MS transmits uplink RLC data blocks during allocated uplink PDTCH occurrence. The network sends a PACKET UPLINK ACK/NACK message when needed on the downlink PACCH. This message indicates the uplink RLC data blocks that have been correctly received and which ones have not been correctly decoded. Figure 5.24 shows a scenario for uplink transfer.

The PACKET UPLINK ACK/NACK message contains the following information:

- *Uplink TFI.* This identifies the uplink TBF.

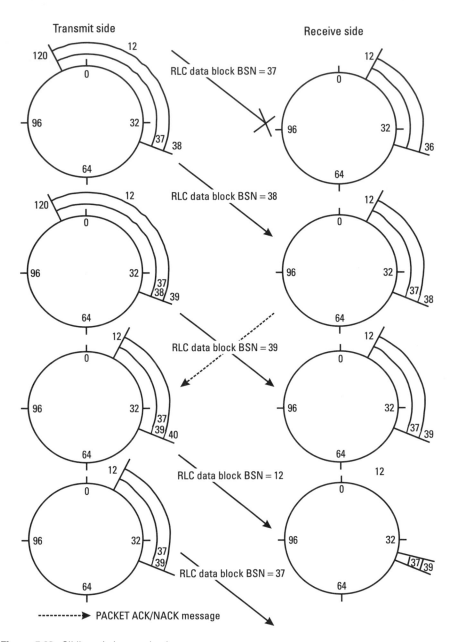

Figure 5.23 Sliding window mechanism.

- *Channel coding command.* The network may request the use of another coding scheme based on measurements performed on the uplink.

- *ACK/NACK description.* This contains three parameters: The FINAL_ACK_INDICATION indicates whether the entire TBF is being acknowledged. It is used for uplink TBF release. The STARTING_SEQUENCE_NUMBER is the BSN corresponding to the end of the window plus 1. The RECEIVE_BLOCK_BITMAP is a bitmap representing whether a BSN is ACK or NACK. The bitmap starts from the STARTING_SEQUENCE_NUMBER backwards.

Note that if fixed allocation is used for the uplink transfer, the BSS may allocate more uplink resources within the PACKET UPLINK ACK/NACK message by including a fixed-allocation bitmap.

RLC Data Block Transfer During Downlink TBF

During downlink transfer, the BSS transmits downlink RLC data blocks to the mobile. Whenever needed, the network requests the sending of a PACKET DOWNLINK ACK/NACK message from the mobile on PACCH.

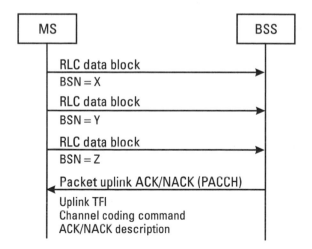

Figure 5.24 RLC data block transfer during an uplink TBF.

The network polls the mobile in an RLC data block in order to request the sending of a PACKET DOWNLINK ACK/NACK message. The polling is performed using the S/P and RRBP fields of the downlink control block MAC header. Figure 5.25 gives a scenario for downlink transfer.

The PACKET DOWNLINK ACK/NACK message contains the following information:

- *Downlink TFI.* This identifies the downlink TBF.
- *ACK/NACK description:* This is the same as for the PACKET UPLINK ACK/NACK message, listed above.
- *Channel quality report.* The network uses these measurements for network-controlled cell reselection and for dynamic CS adaptation.

Note that the mobile may request the establishment of an uplink TBF by including a channel request description within the message.

5.6.3.2 In Unacknowledged RLC Mode

In RLC unacknowledged mode, the sending side transmits the RLC data blocks in sequence. The blocks are numbered so that the receiver is able to detect undecoded data blocks. The data blocks that are not decoded by the receiver are not retransmitted. On the receiver side, when all the RLC data blocks belonging to one LLC frame have been received, the LLC frame is reassembled.

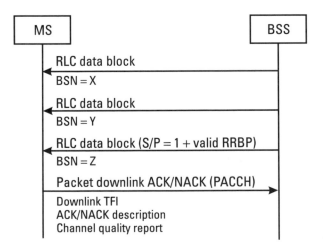

Figure 5.25 RLC data block transfer during a downlink TBF.

During an uplink TBF, the network may send a PACKET UPLINK ACK/NACK message that is used to order a new CS for the mobile or to allocate more uplink resources in case of fixed allocation (in which case a new allocation bitmap is sent to the mobile). During a downlink TBF, the network will request the sending of PACKET DOWNLINK ACK/NACK messages thanks to the polling mechanism in order to receive the channel quality report from the mobile. In spite of the non-retransmission of unde-coded RLC data blocks by the network, the mobile provides the RECEIVE_BLOCK_BITMAP within this message. The network can use this bitmap to estimate the downlink BLER. Note that this evaluation is also possible in RLC acknowledged mode but the evaluation is more complex due to the retransmissions.

5.6.4   Segmentation and Reassembly of RLC/MAC Control Messages

The RLC/MAC control messages contain a lot of optional fields and in some configuration the RLC/MAC message does not fit into one radio block encoded with CS-1. It is possible in the downlink direction to segment one RLC/MAC message into two RLC/MAC control blocks. This possibility is not provided in the uplink direction.

When the network has to transmit one RLC/MAC message within two RLC/MAC control blocks, the optional octet 1 is used to control the transac-tion (see Table 5.2). The RLC/MAC control message is identified by the RTI field of the header. The RBSN field indicates the sequence number of the RLC/MAC control block. The FS field indicates whether the FS of the RLC/MAC control message is contained in the RLC/MAC control block or not.

Note that when the network wants to provide a PR value within the header of a downlink RLC/MAC control block, it must include the optional octet 2 in the header. Or if the message can fit within one radio block, the network must indicate within the optional octet 1 that the message is not segmented. In this case, the network indicates RBSN = 0 and FS = 1.

Figure 5.26 describes an example of PACKET UPLINK ASSIGN-MENT message sending with segmentation.

The network may request the sending of a PACKET CONTROL ACKNOWLEDGMENT message during the sending of the segmented message. The polling is performed using the S/P and RRBP fields of the downlink control block MAC header of both downlink blocks.

By using the CRTL_ACK field, the mobile indicates within the PACKET CONTROL ACKNOWLEDGMENT message whether the two

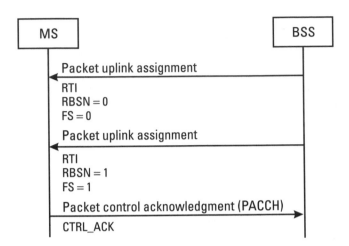

Figure 5.26 Example of RLC/MAC control message segmentation.

radio blocks have been correctly decoded and if not, which one must be retransmitted.

5.7   TBF Release

5.7.1   Release of Uplink TBF

During an uplink transfer, the network allocates uplink resources to the mobile either dynamically with the use of the USF (dynamic allocation or extended dynamic allocation) or at fixed occurrences (fixed allocation). In the case of an open-ended TBF, however, the network does not know exactly when the TBF will end.

In order to avoid the loss of uplink resources allocated by the network (because the mobile has finished its uplink transfer), a countdown procedure has been introduced.

This procedure allows the network to anticipate the end of the TBF and thus to avoid wasting uplink resources. The procedure involves triggering a countdown when the mobile starts sending its last 16 blocks. It uses the CV of the uplink data block header. The procedure is illustrated in Figure 5.27.

When the mobile starts sending its last 16 blocks, it starts decrementing the CV at each transmission. The last block of the TBF is sent with CV = 0.

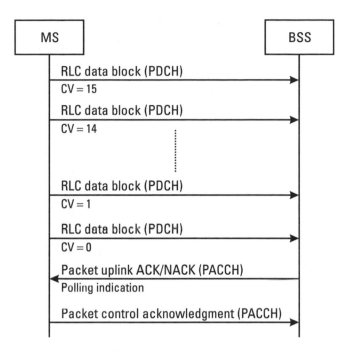

Figure 5.27 Procedure for uplink TBF release.

When the network receives the last uplink RLC data block and if all the blocks have been correctly received, it sends a PACKET UPLINK ACK/ NACK message with the FINAL_ACK_INDICATOR bit set to 1. The network polls the mobile in order to be sure that the mobile has received the PACKET UPLINK ACK/NACK message and to confirm the release of the TBF.

At the reception of the PACKET UPLINK ACK/NACK message, the mobile transmits a PACKET CONTROL ACKNOWLEDGMENT message to the network and releases the TBF.

5.7.2   Release of Downlink TBF

The release of a downlink TBF is simpler. In fact, in this direction, the network directly controls the assignment of downlink resources and can anticipate the end of the transmission. Under this condition, there is no waste of downlink resources. The procedure is described in Figure 5.28.

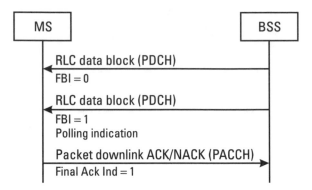

Figure 5.28 Procedure for downlink TBF release.

When the network sends the last RLC data block of a downlink TBF, it sets the FBI to 1 in the last RLC data block (the block that has the highest BSN). This indicates to the mobile the last block of the TBF.

The network requests the sending of the final acknowledgment of the RLC data blocks by polling the MS. In response the mobile sends a PACKET DOWNLINK ACK/NACK message. If all the blocks have been decoded by the mobile, it sets the Final Ack Indicator to 1 within the message.

The mobile continues to monitor its downlink PDCHs during a guard timer. This is to ensure proper recovery in case the PACKET DOWNLINK ACK/NACK message would not be correctly received by the network, in which case the network would resend the last RLC data block with polling (if the network does not receive the PACKET DOWNLINK ACK/NACK message, it cannot know if this is because the PACKET DOWNLINK ACK/ NACK message was lost or because the last RLC data block with polling was lost; therefore, the only possibility for the network is to resend the last RLC data block with polling to the mobile; but of course in order that this proce-dure eventually succeeds, it is necessary that the MS is still listening to the downlink PDCHs; hence the need for the guard timer).

When the network receives the message, it starts a timer and the TBF is released at the end of this timer. However, if in between the network wants to establish a new downlink TBF to the same mobile, it can send a PACKET DOWNLINK ASSIGNMENT message on PACCH as long as the guard timer is running.

5.8   Case Studies

This section deals with some case studies related to practical implementation issues. The first subsection explains which parameters have to be considered when determining the SPLIT_PG_CYCLE value. The second subsection deals with the resource allocation strategy problems. The third subsection addresses the problem of acknowledgment reporting. The last subsection deals with dynamic coding scheme adaptation algorithm.

5.8.1   Determination of SPLIT_PG_CYCLE Value

The goal of this section is to show the impact of the SPLIT_PG_CYCLE value on paging occurrence, on the strategy to perform measurements and BSIC decoding on neighbor cells, and on power saving management for the MS. The parameter SPLIT_PG_CYCLE indicates the paging occurrence of a paging subchannel within a 52-multiframe or 51-multiframe in DRX mode for GPRS services. This parameter is positioned by the MS during the GPRS attach procedure and is not negotiated between the MS and the network.

The SPLIT_PG_CYCLE directly impacts the following actions performed by the MS in packet idle mode:

- Paging decoding according to DRX parameters;
- Measurements on the serving cell and on neighbor cells for the cell-reselection algorithm;
- BSIC decoding on each of the six best neighbor cells.

As will be shown in the next sections, the paging recurrence has an impact on the two last actions listed above. This case study concerns SPLIT_PG_CYCLE only for the 52-multiframe, not the 51-multiframe.

5.8.1.1   Performed Measurements

There is a benefit for the MS in performing all measurements for cell reselection during the paging decoding process in order to stop its radio activity between two consecutive paging blocks. This operation allows for a limiting of the battery consumption in idle or packet idle mode.

Note that between two paging periods, the MS enters a low-power mode where the radio activity is fully stopped. As waking up of the radio activity is not instantaneous, the MS enters a normal power-management mode a very short time before the paging block decoding. As there is a latency time for radio activation, there is a benefit for the MS in limiting the number of radio activations for cell-reselection measurements in order to optimize battery consumption. That is the reason that it is good to take advantage of the paging-decoding process for cell-reselection measurements.

Measurement Window

As a PPCH block consists of four bursts on four consecutive TDMA frames, the MS may open measurement windows on time slots that are not used for paging block decoding. For example, with a quickly configurable and reactive radio, it will be possible to open several measurement windows within a TDMA frame. The number of measurement windows for a given MS is limited by two factors (see Figure 5.29):

- Time to perform a measurement on a carrier;
- Time to lock on a new frequency.

GPRS Constraints

The MS is required to calculate a running average on measurements collected on a period equal to MAX (5s, five consecutive paging blocks) in order to evaluate the cell-reselection criteria. At least five RXLEV measurements are taken for each carrier. The network broadcasts a list of neighbor cells,

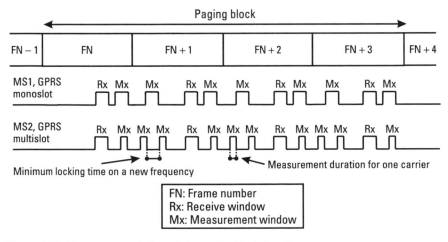

Figure 5.29 Measurement windows during paging block decoding.

called the BA list (BCCH allocation). This list may contain up to 32 neighbor cells, meaning that the MS will perform up to 5×33 measurements on neighbor cells plus the serving cell during a period equal to MAX (5s, five consecutive paging blocks).

The occurrence of a paging block within a 52-multiframe according to DRX parameters is defined by:

$$64/\text{SPLIT\_PG\_CYCLE} \times 0.240\text{s} \qquad (5.28)$$

where 0.240s is the time duration of a 52-multiframe.

As the range of SPLIT_PG_CYCLE values is between 1 and 352, the occurrence of a paging block lies between 0.0436s and 15.36s. When SPLIT_PG_CYCLE is greater or equal to 16, the paging occurrence becomes lower than 1s. Thus the average running for cell reselection measurements is equal to 5s for all these SPLIT_PG_CYCLE values. Table 5.10 gives the period for the running average.

Moreover, the recommendation gives the following constraints:

- At least one measurement on each neighbor cell every 4 seconds;
- One measurement as a maximum on each BA list carrier every second.

SPLIT_PG_CYCLE Value Less Than or Equal to 3 As the MS will respect the constraint of at least one measurement on a carrier every four seconds, the opening of a measurement window occurring every five consecutive paging blocks is not enough for these SPLIT_PG_CYCLE values (see Table 5.10). In this configuration, the MS will open additional measurement windows outside the ones opened during paging block decoding in order that the constraint described above be respected.

SPLIT_PG_CYCLE Values Less Than 16 and Greater Than 3 As the measurement period is equal to "5 five consecutive paging blocks" and as five RXLEV measurements are taken for each carrier, the MS will perform one measurement on each neighbor cell belonging to the BA list plus on the serving cell. If the number of neighbor cells plus the serving cell is greater than the number of measurement windows opened during paging block decoding, then the MS will open additional measurement windows outside the TDMA frames containing paging blocks.

Table 5.10

Cell Reselection Measurement Periods

SPLIT_PG_CYCLE	Period for Running Average (s)
1	76.80
2	38.4
3	25.6
4	19.2
5	15.36
6	12.8
7	10.9714
8	9.6
9	8.5333
10	7.68
11	6.9818
12	6.4
13	5.9077
14	5.4857
15	5.12
16	5
> 16	5

SPLIT_PG_CYCLE Values Greater Than or Equal to 16   As the measurement period is equal to 5s, there is a number n of paging blocks during this period (e.g., for SPLIT_PG_CYCLE = 16, n = 5) and since the MS is able to perform $n \times N$ measurements on $n \times N/5$ cells (neighbor cells plus serving cell) during this period, the number N represents the number of measurement windows opened during paging block decoding. If the number of neighbor cells is greater than or equal to $n \times N/5$, then the MS will open additional measurement windows outside the ones opened during paging block decoding. If the number of neighbor cells is lower than $n \times N/5$, then the MS will not be obliged to open all $n \times N/5$ measurement windows during 5s period in order to save battery consumption.

Note The SPLIT_PG_CYCLE is a fixed value in the MS; it does not change according to the number of neighbor cells. This means that it makes sense to choose a SPLIT_PG_CYCLE value that allows measurements to be performed on the maximum number of neighbor cells (32) during five seconds without opening additional windows outside the TDMA frames containing paging blocks.

5.8.1.2 BSIC Decoding on Neighbor Cells

The MS is required to check the BSIC on each of the six best neighbor cells at least every 14 consecutive paging blocks or 10s. The BSIC decoding process is performed outside the paging block decoding process.

Note The period of 14 consecutive paging blocks depends on the SPLIT_PG_CYCLE value. For a SPLIT_PG_CYCLE value greater than 22, the BSIC decoding is performed every 10 seconds; otherwise it is performed every 14 consecutive paging blocks.

5.8.1.3 Choice of SPLIT_PG_CYCLE Value

As discussed above, the SPLIT_PG_CYCLE value has an impact on the following points:

- Paging reactivity: range between 0,0436s and 15,36s;
- Serving and neighbor-cells measurement strategy;
- BSIC decoding on neighbor cells;
- Power-saving management.

A high value for SPLIT_PG_CYCLE is required for a good paging reactivity. In contrast, a low value for SPLIT_PG_CYCLE is beneficial in terms of power-saving management. A very low value for SPLIT_PG_CYCLE is risky for the MS, since it may lose synchronization in frequency and in time with the network if the paging occurrence is too long; the MS uses the paging block decoding to adjust frequency and time drift. The MS makes a tradeoff between these criteria, then, in order to determine a SPLIT_PG_CYCLE value.

Note that a value for SPLIT_PG_CYCLE lying between 8 and 32 seems a good tradeoff between paging reactivity and power savings. This value corresponds to a BS_PA_MFRMS value included between 2 and 8 used for paging recurrence on the 51-multiframe.

5.8.2   Resource Allocation Strategy

5.8.2.1   Sharing of Time Slots Between Circuit-Switched and Packet-Switched Services

From a radio point of view, the evolution of a GSM network toward GPRS requires the introduction of radio resources dedicated to packet transfer within the cells. This is the case in most of the networks, as most of them will support both packet- and circuit-switched modes. PDCHs are introduced within cells already having BCCH, SDCCH, TCH, and so on.

The introduction of GPRS dedicated radio resources has a direct impact on available circuit-switched resources. The more PDCHs available in the cell, the less TCHs if GPRS is introduced without any change in the radio network planning. In order to introduce GPRS as smoothly as possible and avoid impacting speech service, it is important to optimize the allocation of PDCHs within the cell and have a coordinated strategy between packet and circuit radio resource management.

The RR allocation algorithm will also be able to manage the increase of packet traffic with time. At the introduction of GPRS service within networks, in order not to impact the speech service, an operator will probably request the allocation of few PDCHs within the cell. The allocation of PDCHs should be dynamic and reactive due to the small number of GPRS users in the introductory phase of the system.

In the same way, when the load of speech calls increases within a cell, it is important to keep enough free radio resources within the cell in order that incoming speech calls are able to be accepted in the cell. These incoming speech calls can be due to a mobile trying to access the network or to handovers from other cells. In fact, from a user point of view it is probably more important to maintain a speech call (successful handover) than to maintain packet transfers.

Thus in order to avoid an increase of the speech-call blocking rate, operators may request dynamic allocation of PDCHs depending on the circuit-switched load and packet load.

When the number of GPRS users increases, it will become necessary to have permanent GPRS radio resources in the cell, so that GPRS users are very quickly allocated resources.

In order to take advantage of the multislot class of the mobile, the network must optimize the allocation of PDCHs. In fact, if two PDCHs are allocated in the cell on two different frequency allocations, it will not be possible to allocate the two PDCHs to the same mobile during one TBF. Furthermore, if the two PDCHs are not consecutive, it will not be possible to allocate two PDCHs (even if they have the same frequency allocation) to a class 4 mobile, even if it can support two received time slots.

In order to avoid this kind of problem, the network must allocate PDCHs in a very efficient way. In order to increase the probability of consecutively allocated PDCHs, it could be possible on the network side to reserve a pool of radio resources for circuit-switched only, another one to packet services only, and a last one allocated to both packet and circuit. Figure 5.30 gives an example of a possible mapping between these different pools. In this example, the cell supports three hardware units, each one handling reception and transmission (TRX) on a different frequency allocation. Up to eight TCHs can be supported on TRX 2.

The resource allocation algorithm can then be as follows: When there is an incoming speech call, it is allocated a resource in the circuit-switched pool when available. This avoids monopolization of a resource in the common pool that could be allocated to another GPRS user during the speech call. When there is an incoming speech call and no resource is available in the circuit-switched pool, it is allocated a resource in the common pool. However, it is very important in this pool to avoid a random mixing of TCHs and PDCHs in order to avoid breaking up consecutive resources that can be used for PDCHs. In our example, a solution to this problem could be

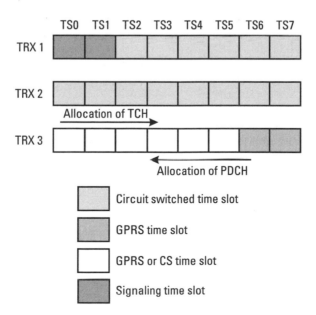

Figure 5.30 Example of time slot allocation in a cell.

to allocate PDCHs from the right to the left and TCHs from the left to the right on TRX 3.

The last case is when a TCH is requested and no resources are available in both pools. Either the speech call is rejected or it could be possible to release a PDCH and allocate a TCH in place of this latter. In the same way, the allocation of a PDCH is done in the packet pool if resources are available, or otherwise in the common pool.

This allocation mechanism is very flexible, as it also allows for a smooth introduction of packet resources within the network. The only constraint for the operator will be to choose the right size for the different pools. At the introduction of the GPRS service, the strategy will be to have a small packet pool, and when the GPRS traffic has grown, to increase the size of this pool.

Another key issue when allocating PDCHs is the average throughput that will be available. As described in Chapter 4, during a TBF, link adaptation is used to adapt the throughput on the channel depending on the quality of the radio link. So the greater the average C/I for a TRX, the higher the average throughput on the PDCHs of this TRX. In most of the networks today, the average C/I is around 11 or 12 dB. This corresponds to an average throughput per PDCH of 12 Kbps. A user that is near the BTS will get a higher throughput but a user far from the BTS will get a lower one.

However, an operator that wants to provide a better GPRS service with higher average throughputs could decide to have a dedicated frequency planning for TRXs that support GPRS (in our example, TRX 3). Another possibility will be to map the packet pool on the TRX that supports BCCH, as this one is generally well planned (highest C/I) because it supports BCCHs.

This section has given an example of an algorithm for the allocation of PDCHs and TCHs at the network side. The intention was to show the considerations that must be taken into account from a BSS implementation point of view and the impact of the GPRS introduction of RR management.

5.8.2.2   Dynamic Allocation of PBCCH and PCCCH

As the presence of PBCCH and PCCCH is not mandatory within a cell, GPRS access can be performed via CCCH when PBCCH is not present in the cell or PCCCH when PBCCH is present. When PBCCH is not present in the cell and GPRS service is supported, all the mobiles within the cell access the network on CCCH.

The introduction of GPRS mobiles in the network will considerably increase the load on these CCCHs. In fact, a packet transfer such as a Web page download can require a lot of TBF establishments (e.g., request, answer, acknowledgment) within a few seconds' duration. As soon as the GPRS traffic in the cell increases, the common channels will be more loaded and a high risk of congestion will appear, impacting the circuit-switched access and degrading the speech service.

In this case, the allocation of a PDCH carrying PBCCH/PCCCH becomes necessary to handle the increasing flow of GPRS access. The allocation of the PBCCH/PCCCH can be either static (one or more PBCCH/PCCCH are statically allocated within the cell) or dynamic (depending on the load on CCCH or PCCCH, one or more PBCCH/PCCCH are allocated).

Moreover, when the GPRS load in the cell decreases, these PDCHs can be deallocated.

The dynamic allocation of PBCCH/PCCCH can take into account the following load indicator:

- Load on RACH/PRACH;
- PRACH collision indicating the number of collisions detected on RACH or PRACH;
- Load on PPCH;
- Load on PCCCH;
- Evaluate the rate of successful access at the first attempt using the retry bit of the RLC/MAC header.

Depending on these measurements, the network can evaluate whether the allocation of a PBCCH/PCCCH is necessary and whether additional PCCCHs are necessary in the cell.

5.8.2.3 Allocation of MS Radio Resources

PDCH Allocation

This section describes the constraints that have to be taken into account by the network in allocating PDCHs to an MS.

In case of a downlink TBF, the network allocates downlink PDCHs and one uplink PDCH that carries the uplink PACCH. For an uplink TBF, when fixed allocation is used, the network allocates uplink PDCHs and one downlink PDCH for the downlink PACCH. When dynamic allocation is used, the network allocates the same number of PDCHs in uplink and downlink, since the transmission on one uplink PDCH is allowed by the decoding of an allocated USF on the associated downlink PDCH.

The information that can be used by the network to allocate the PDCHs is the multislot class of the mobile when it is available, and the type of TBF (uplink or downlink) and its nature (GMM procedure signaling or data transfer). The number of PDCHs that can be allocated to the mobile is dependent on its multislot class and the availability of PDCHs in the cell.

The network is interested in allocating as many PDCHs matching the multislot class of the mobile as possible, in order to provide a higher throughput and increase the multiplexing gain. However, this is not for all the TBFs. In fact, for example, TBFs that are used to carry GMM signaling do not need to be allocated multiple PDCHs, since from a user point of view there is no difference if this kind of TBF is carried with a high throughput or not. It can be handled on a single PDCH.

The network must also take into account the number of mobiles that are multiplexed on the same PDCHs. If too many TBFs are allocated on the same PDCH, the available bandwidth per mobile will be very low and the experienced service very bad for all the users.

One problem inherent in GPRS is the allocation of the uplink PDCH. Due to the multislot constraints, the allocation of uplink PDCH is often shifted on the highest TN as compared with the downlink allocation. Figure 5.31 shows the possible PDCHs mapping for multislot class 6 and class 12 mobiles during downlink transfer. For the class 6 mobile, TS 5 and TS 7 cannot be allocated. For multislot class 12, time slots 2, 3, and 5 will never be allocated in uplink.

The consequence of this is that the uplink traffic is not well distributed between all the available uplink time slots. However, we can consider this to be a minor problem, being that the downlink traffic is much higher than the uplinktraffic.

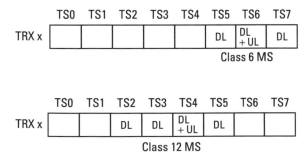

Figure 5.31 Mapping of class 6 and class 12 mobiles on a TRX.

Figure 5.32 Mapping of a class 6 mobile on a TRX when the uplink time slot is at the border of the TRX.

However, when the network allocates uplink PDCHs to an MS, it will take care not to allocate the uplink resources on the extreme side of the PDCH pool on a TRX. For example, the mobile requests the allocation of an uplink TBF and the network allocates a PDCH on the extreme left of the TRX. If a concurrent downlink TBF is established during the uplink transfer, the BSS will not be able to allocate in downlink as many PDCHs as could be supported by the multislot class. It could be possible to change the uplink allocation of the mobile, but this will make the handling of resources more complex.

Figure 5.32 shows that for a class 6 mobile, if an uplink TBF is handled on time slot 0, the PDCHs allocated to the concurrent TBF can only be mapped on time slot 0 and 1, although three time slots could have been supported in downlink.

Multiplexing of Mobiles onto PDCHs

Once PDCHs have been allocated to the different mobiles for uplink or downlink transfers, the network must manage the allocation of downlink and uplink blocks. Since different users can be multiplexed on the same PDCHs, the BSS must share the bandwidth between the different mobiles.

Different strategies can be considered for the allocation of blocks when a PDCH is shared between several mobiles. One can be the allocation of the full PDCH to a mobile until its transfer is finished, and then allocation of the PDCH to the next TBF. The major drawback of this solution is that it delays the other TBFs. When the current TBF is very long, the other TBFs cannot be allocated resources on the PDCH. The other solution is to share alternatively and dynamically the bandwidth on the PDCH between the different users that are multiplexed on it.

Downlink Multiplexing Figure 5.33 shows how the downlink multiplexing can be managed at the network side. In this example, three PDCHs are allocated on the TRX. The first PDCH on the left supports PBCCH and PCCCH; the two others are used for data transfer. The PDCH that supports the

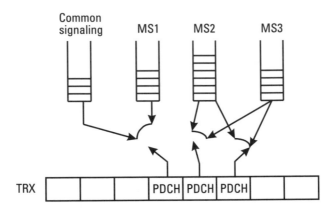

Figure 5.33 Example of downlink multiplexing.

PBCCH is allocated to MS1. MS2 and MS3 are multiplexed on the two other PDCHs. One queue that contains the segmented RLC data blocks is dedicated to each mobile.

If the algorithm consists of giving alternatively downlink block allocations to the mobiles, on the second PDCH the first block period will be allocated to MS2, the second to MS3, the third again to MS2, and so on. The multiplexing of the PDCH that supports PBCCH will be exactly the same, except that for PBCCH and PCCCH fixed occurrences signaling must be transmitted.

Depending on the QoS of the mobile, the network may decide to give more or less consecutive block occurrences to a mobile on a given PDCH. If the QoS of MS2 is higher, MAC may allocate three block occurrences to MS2 and one to MS1 alternatively.

Note   The downlink scheduler must also manage the transfer of signaling blocks for the uplink TBFs that share the same PDCH.

Uplink Multiplexing   For the uplink, the management of the multiplexing is quite different depending on whether the network uses dynamic allocation or fixed allocation. Dynamic allocation allows uplink block occurrences to be allocated with a very short anticipation. With this kind of multiplexing scheme, the same principle as for the downlink can be reused. The USFs can be distributed to the different mobiles that share the same PDCH alternatively. This ensures the sharing of the uplink bandwidth. However, the scheduler must also manage the fixed-block instance reserved by the RRBP values scheduled in downlink blocks. The scheduler during these instances will avoid scheduling a USF value that is allocated to a mobile on this PDCH in order to avoid collision on the uplink PDTCH occurrence.

When the network uses fixed allocation, the sharing of the bandwidth between the different mobiles is not as easy as for the dynamic allocation scheme. For this mode, the network allocates many uplink block occurrences to the mobile with anticipation (bitmap mechanism).

If two mobiles are sharing the same PDCH, the network distributes the bandwidth between the two mobiles, allocating one part of the bandwidth to one mobile and the other part to the other one. But if during the uplink transfer a new mobile is allocated the same PDCH, the network will need to reserve some bandwidth for it. Moreover, when there are downlink TBFs sharing this PDCH, some uplink bandwidth must be reserved for the transmission of their uplink PACCH.

This management of bandwidth allowing for other mobiles is very difficult to handle at the network side. It is difficult to forecast how much bandwidth will be required for downlink TBF signaling and when it will be needed. One possible solution would have been to reduce the length of the fixed resource allocation. This could lend more flexibility to the allocation scheme, but it requires a frequent sending of resource allocation messages, bringing about a reduction of the downlink bandwidth for data. The other solution involves regularly reserving some bandwidth for unexpected uplink block occurrences. The drawback of this solution is that, when this additional bandwidth is not requested, it cannot be allocated to the uplink users due to the slow reaction time of the fixed allocation scheme.

Figure 5.34 shows how fixed allocation can be managed at the network side with an uplink allocation table. Each row of the table stands for a PDCH, and the columns for block periods. On the left of the table, the mobiles that are multiplexed on each PDCH are indicated. Each block period on a PDCH is reserved either for one mobile, for signaling (PBCCH,

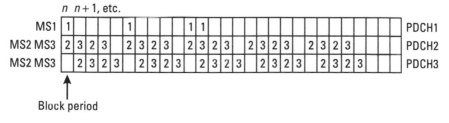

Figure 5.34 Example of resource management for the fixed-allocation multiplexing scheme.

PCCCH) or for other unexpected needed uplink occurrences. Each time a block period occurs on the air interface, the corresponding column is removed from the table.

5.8.3   ACK/NACK Request Period Within RLC Layer

In Section 5.6 we described the RLC protocol based on the sliding window mechanism. When operating in RLC acknowledged mode, this protocol manages the retransmission of incorrectly decoded RLC data blocks thanks to an acknowledgment bitmap.

As previously explained, the sending side transmits blocks as long as the end of the transmit window is not reached. Once the end of the window is reached, the transmit side is not allowed to transmit new RLC data blocks. It is only allowed to retransmit RLC data blocks that have not yet been acknowledged. The RLC protocol is stalled; the transfer of data no longer progresses.

The task that manages the request for the sending of an acknowledgment message is located in the BSS. In the case of uplink transfer, it sends PACKET UPLINK ACK/NACK messages that acknowledge the correctly decoded uplink blocks. During downlink transfer it orders the sending of PACKET DOWNLINK ACK/NACK messages from the mobile.

For both TBF types, the main goal of this task is to avoid having the RLC protocol stall. It must send (or request, in case of downlink transfer) acknowledgment messages at a sufficient rate to avoid the RLC protocol stalling. But the number of acknowledgment messages must not be too high, in order to avoid overloading the PDCH that carries the PACCH for acknowledgment messages. Since this PDCH can be shared with other mobiles, it would reduce the bandwidth available for the other mobiles. The algorithm must reach a good compromise between PACCH load and RLC stall avoidance.

5.8.3.1   Downlink TBF

For a downlink TBF, the network requests the sending of the PACKET DOWNLINK ACK/NACK message from the mobile by including a valid RRBP value in the header of the downlink RLC data block. The PACKET DOWNLINK ACK/NACK message is sent by the mobile some block periods after, as specified by the RRBP field. This field specifies a single uplink block occurrence by giving the number of TDMA frames the mobile must wait before transmitting. Four values are available $(N + 13)$, $(N + 17$ or $N +$

18), (N + 21 or N + 22), or (N + 26) where N is the frame number in which is received the first burst of the block giving the RRBP value. In two cases, two values are given for example (N + 17 or N + 18). N + 17 corresponds to the case where there is only one idle frame between the reception and the transmission; in N + 18 there are two idle frames.

There is some delay between the request of the ACK/NACK message and its reception at the network side. This delay is linked to the RRBP period and to the global round-trip delay. The round-trip delay is for example the time that elapses between the sending of a USF value within one downlink radio block and the reception of the answer at network side. This time is dependent on the PCU location within the BSS. The closer the PCU to the BTS, the shorter the round-trip delay. Depending on the BSS implementation, the delay between the ACK/NACK message request and its reception can vary from 100 ms and 300 ms. This corresponds to 5 to 15 block periods.

This means that if a mobile is assigned one PDCH and it is not multiplexed with another mobile, between the request of the ACK/NACK message and its reception, the network will have sent between 5 to 15 more downlink data blocks to the mobile depending on the round-trip delay. Now, if we consider a mobile that is allocated 4 PDCHs, it will receive four times more blocks during this period, equivalent to 20–60 downlink data blocks.

In case of 4 time slots, the number of blocks that can be received during the polling period can be very important. The RLC window of 64 can be filled just between the request and the reception of the acknowledgment message. In the case of 4 time slots and considering a round-trip delay of 300 ms, it will be necessary to poll the mobile for the next acknowledgment message transmission before having received the first ACK/NACK message.

One possible solution for the polling algorithm is to poll the mobile after having addressed to it a constant number of blocks (for example, the mobile is polled every 20 blocks addressed to it). This number of blocks can be adapted dynamically depending on the state of the sending window. When the RLC protocol is about to be stalled, the polling period is reduced in order to receive ACK/NACK messages more often.

Figure 5.35 describes the way the polling mechanism works. In this example, the network requests the sending of an ACK/NACK message every x blocks sent. The first polling indication is sent within the block with BSN = x − 1. So the downlink ACK/NACK is received after the BSS has sent block y. Then the network retransmits unacknowledged downlink RLC data

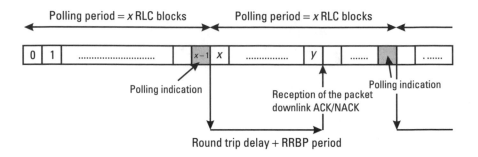

Figure 5.35 Polling mechanism.

blocks. The second polling indication is sent after the $2 \times x$ block transmitted by the BSS.

Note that the choice of the RRBP value by the network at the moment it wants to poll a mobile is dependent on the availability of an uplink block occurrence. The network chooses the minimum RRBP value available in order to reduce the round-trip delay. However, some of the RRBP values may be unavailable, since multiple mobiles can be multiplexed on the same PDCH and some uplink block occurrences have been reserved for those mobiles limiting the choice in the RRBP value.

5.8.3.2   Uplink TBF

For an uplink TBF, the receiver is the BSS. It triggers the sending of PACKET UPLINK ACK/NACK messages indicating to the mobile which blocks have been correctly received. For this kind of TBF, no round-trip delay takes place in the process of uplink block acknowledgment. As soon as the network receives a block, it knows the difference between the oldest unacknowledged block and the block decoded with the highest BSN. If this difference in terms of BSN reaches 64, the protocol is stalled.

The network can easily avoid this situation, as at the reception of each block it knows whether or not the RLC protocol at the mobile side is about to stall. If this is indeed the case, it can decide to send an uplink acknowledgment message.

The same mechanism as for downlink transfer can be used. The network sends PACKET UPLINK ACK/NACK messages regularly, for example every x received RLC data blocks. The value of x can be larger than that used for the downlink transfer, since the influence of the round-trip delay does not have to be taken into account (there is only a half round-trip delay).

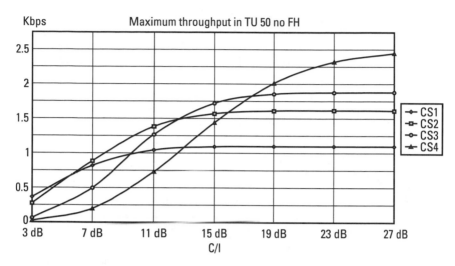

Figure 5.36 Throughput versus C/I for the different coding schemes (TU50 no FH). (*From:* [1].)

5.8.4 Implementation of Dynamic Link Adaptation

As seen in Section 4.1.2.1, the coding rate is dynamically adapted according to the radio conditions, with a tradeoff between protection and achieved throughput. This section presents the different parameters that must be taken into account in the link adaptation mechanism and the different possibilities of algorithm implementation.

Figure 5.36 shows the maximum throughput versus C/I for a mobile moving at 50 km/hr in a *typical urban* (TU) environment in a cell where no frequency hopping is used. The results for the different coding schemes have been drawn. The results are obtained with a tool that simulates the radio link and the reception chain. The upper layer constraints are not taken into account (there is no retransmission at RLC layer, no link adaptation, no signaling). Simulations are performed for different average C/I values. The throughput is computed as the number of correctly decoded blocks over the total number of blocks that have been transmitted.

An ideal link adaptation would always choose the coding scheme giving the maximum throughput for an instantaneous C/I value. However, in reality, the link adaptation algorithm must take into account a lot of practical parameters, as follows:

- The C/I evaluation is not perfect, and will be averaged in order to prevent decisions on a single instantaneous value that does not

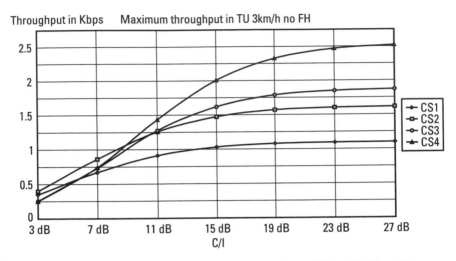

Figure 5.37 Throughput versus C/I for the different coding schemes (TU3 no FH). (*From:* [1].)

reflect the actual usability of the link for communication but only the particular propagation conditions at one instant in time.

- In case of uplink transfer, as the network commands the MS to use a particular coding scheme in uplink, there is a delay between the decision and the usage of the new coding scheme by the MS. During this delay, the radio condition could have changed, which means that some margins have to be taken around the switching points.

- The switching point between the different coding schemes is not the same depending on the mobile speed, the environment (TU, rural area, and hilly terrain) and the usage or not of frequency hopping (see Figure 5.37).

- As the transmission mode is packet oriented and the same medium is shared between different mobiles, the network does not continuously monitor the link quality for a given MS.

5.8.4.1 Link Adaptation Metrics

In order to adapt and choose the most adequate coding scheme, the network can use different measurements.

Downlink Adaptation

During downlink transfer, the MS measures the average quality (RXQUAL), the average RXLEV (C value) and the interference level on the different time slots. These measurements are periodically reported to the network following its request.

The network can evaluate the average C/I based on the average C level and the interference level received, and depending on the C/I value determine the most appropriate coding scheme.

Another possibility is to use the average RXQUAL as the switching condition between the different coding schemes. The average RXQUAL corresponds to the BER before channel decoding; thus it reflects more or less the average C/I. However, when the network transmits with CS-4, the mobile is unable to estimate the RXQUAL (no convolutional coding is used for CS-4); so for CS-4, another criterion must be used by the network. The criterion in this case could be the C/I ratio.

There is also the option of using an evaluation of the BLER as a criterion for switching between the different coding schemes. In fact, this parameter is directly linked to the average throughput. Nevertheless, this information is not fully reliable, if it is not used with another metric, as the MS can sometimes not monitor its downlink PDCH due to the neighbor cell information decoding, so the BLER in this case could be overestimated.

Uplink Adaptation

For the uplink, the network can evaluate the quality of the link using the average RXQUAL as described for the downlink, or it can evaluate the C/I based on the training sequence.

5.8.4.2 Link Adaptation Algorithm

One simple algorithm involves defining three fixed C/I thresholds defining the switching boundary between the different coding schemes. When the average C/I is higher than the threshold plus a margin, a less-protected coding scheme is chosen. When the average C/I falls below the threshold minus a margin, a more robust coding scheme is used. The margin is used as a hysteresis to avoid a ping-pong effect from one coding scheme to the other around the boundary. A more sophisticated algorithm could involve dynamic thresholds.

As stated previously, the switching point between two coding schemes is dependent on the mobile environment (TU, rural area, and hilly terrain). Knowing this, it is possible to estimate the environment in which the mobile is and then adapt the threshold depending on the environment.

Note that the BLER can be computed in uplink in the following way. For each block received at the network side, the BTS checks the BCS of the block to determine whether or not the block has been correctly decoded. The RLC/MAC layer, knowing to which mobile it has assigned the uplink resource on which the block was received and decoded or not, can deduce the BLER relative to this mobile. In the downlink direction, the network can compute the BLER based on the acknowledgments that are reported by the mobile.

An important issue when dealing with the link adaptation is its interaction with the power control algorithm. In fact, reducing the transmission level in uplink or in downlink impacts the quality of the transmission. Reducing the power has the effect of lowering the throughput. However, reducing the transmission power decreases the global interference level in the network and improves the transmission quality. Thus a tradeoff between PR and throughput must be found, although this is not necessarily easy to do.

5.8.4.3 Initial Coding Scheme Determination

At the beginning of the packet transfer, the BSS has no knowledge of the link quality between the BTS and the mobile, so there is no real metric allowing the determination of the initial coding scheme. One method is to always use a robust coding scheme (CS-1 or CS-2) at the beginning of the TBF in order to begin the TBF safely. Once a first evaluation of the link quality is available, the network can start to adapt dynamically the coding scheme.

However, between two TBFs, the network can keep in memory the last coding scheme used by the mobile during the previous TBF and reuse it as the initial coding scheme for the next TBF, as long as the time between the two TBFs is not too long.

Reference

[1]   3GPP TS 05.50 Background for Radio Frequency (RF) Requirements (R99).

Selected Bibliography

3GPP TS 03.64 Overall Description of the GPRS Radio Interface; Stage 2 (R99).

3GPP TS 04.18 Radio Resource Control Protocol (R99).

3GPP TS 04.60 Radio Link Control/Medium Access Control (RLC/MAC) Protocol (R99).

3GPP TS 05.08 Radio Subsystem Link Control (R99).

6

Gb Interface

This chapter covers signaling and data transfer over the Gb interface. This interface is essential, as it is the link between the GPRS core network and the BSS. It is used for the transport of GMM, *packet flow management* (PFM), and NM signaling.

Section 6.1 provides an overview of the Gb interface with its associated protocol stack and the services provided by the different layers. Section 6.2 covers the frame relay protocol, used as a transport network over the Gb interface. Section 6.3 explains the different levels of addressing between the SGSN and the BSS. Section 6.4 details the NS layer, and Section 6.5 deals with the BSSGP layer. Section 6.6 offers case studies showing how the different procedures interact in some typical situations, such as cell reselection.

6.1 General Overview

Figure 6.1 shows the position of the Gb interface within the GPRS network. The Gb interface is located between the BSS and the SGSN (between the gray boxes in Figure 6.1). It allows for the exchange of signaling information and user data, and the multiplexing of many users on the same physical resource. This interface is completely standardized, allowing interworking between the BSSs and SGSNs provided by different manufacturers.

Figure 6.2 shows the protocol stack between the SGSN and the BSS. The NS layer has been split into two sublayers in order to have the NSC sub-

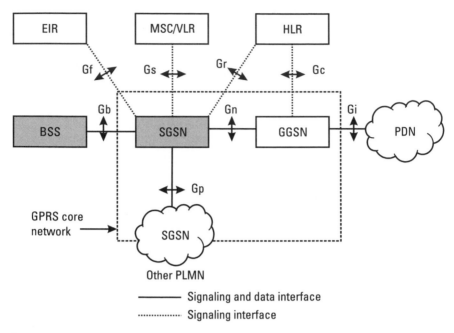

Figure 6.1   Position of the Gb interface.

layer independent of the intermediate transmission network, which is based at the moment on the frame relay protocol.

The peer-to-peer communication across the Gb interface between the two remote NS entities in the BSS and the SGSN is performed over virtual

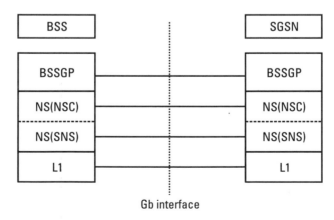

Figure 6.2   Protocol stack on the Gb interface.

connections. The NS layer is responsible for the management of the virtual connections between the BSS and the SGSN (verification of the availability of the virtual connections, initialization, and restoring of a virtual connection). It ensures the distribution of upper-layer PDUs between the different possible virtual connections (load-sharing function).

The BSSGP layer ensures the transmission of upper-layer data (LLC PDUs) from the BSS to the SGSN or from the SGSN to the BSS. It ensures the transmission of GMM, PFM, and NM signaling. The peer-to-peer communication across the Gb interface between the two remote BSSGP entities is performed over virtual connections. There is one virtual connection per cell.

6.2   Frame Relay Basics

As mentioned previously, the NS layer relies on the frame relay protocol. This section provides a brief background discussion of the protocol. Frame relay is a common protocol that is used in many packet-switched networks.

Frame relay is an evolution of the X25 packet-switched technology. It is based on the concept of *virtual circuits* (VCs). There are two types of VCs: *permanent virtual circuits* (PVCs) and those allocated on demand, or *switched virtual circuits* (SVCs).

The Gb interface relies on the allocation of PVCs only.

PVCs are set up by the network operator via a NM system. They are defined as a connection between two endpoints. They are fixed paths; this means that they are not allocated dynamically or on a per-call basis. Unlike X25, frame relay eliminates all layer 3 processing. Only a few layer 2 functions are used, such as checking for a valid, error-free frame but not requesting transmission if an error is found.

Note that all layer 3 processing has been eliminated due to progress in the reliability of transmission means. As the transmission is extremely reliable, only very few frames are received erroneously; when a frame relay switch receives an erroneous frame, it discards it.

6.2.1   Frame Format

Frame relay uses a variable-length framing structure. This structure is shown in Figure 6.3. Two flags are delimiting the frame. A *frame check sequence* (FCS) is added for integrity protection.

The address field DLCI has a variable length (6- or 10-bit length). The *end address* (EA) is used to manage the length of the address field. When it is

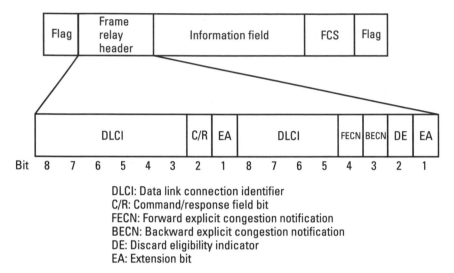

DLCI: Data link connection identifier
C/R: Command/response field bit
FECN: Forward explicit congestion notification
BECN: Backward explicit congestion notification
DE: Discard eligibility indicator
EA: Extension bit

Figure 6.3   The FR frame format.

set to 0, the next octet contains the rest of the address; otherwise, it indicates the last field of the address. The C/R bit indicates whether the frame is a command frame or a response frame.

The BECN and FECN bits are used to avoid congestion. They are used when a congestion situation is about to occur in one direction or in the reverse direction. The source endpoint that receives this information will reduce its throughput.

The DE bit allows the network nodes to indicate which frames will be eliminated first in case of congestion, as described in Section 6.2.3.

6.2.2   Addressing

The frame relay header contains a 10-bit field called the *data link connection identifier* (DLCI). The DLCI is the frame relay VC number (with local significance) that corresponds to a particular destination. It is a number used to distinguish the connection between the endpoint and the frame relay switch from all other connections that share the same physical port.

The DLCI allows data coming into a frame relay switch to be sent across the network using a simple three-step process: First, the integrity of the frame is checked using the FCS. If an error is detected, the frame is discarded (error recovery is performed at a higher level). The incoming DLCI is then looked up in the routing table. The associated outgoing link is identi-

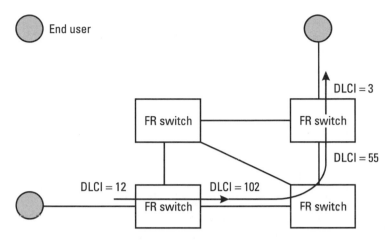

Figure 6.4 Example of FR network.

fied, together with the outgoing DLCI. The frame is relayed toward its destination by sending on the link specified in the table with the outgoing DLCI.

Figure 6.4 illustrates a frame relay network composed of four FR switches. Between the two end users, a PVC is established. The arrow shows the commutation in each switch and the DLCI associated with each link that is used to define the PVC. DLCI numbers change through the network from switch to switch for the same PVC. They have only local significance for each port.

6.2.3 Flow Control Mechanism

In severe congestion, the overall network throughput can diminish, and the only way to recover is for the user endpoints to reduce their traffic. For this reason several mechanisms have been developed to notify the user endpoints that congestion is occurring and that they should reduce their offered load.

There are two types of mechanisms to minimize, detect, and recover from congestion situations:

1. Explicit congestion notification;
2. Discard eligibility.

These mechanisms use specific bits contained within the header of each frame. The locations of these specific bits (FECN, BECN, DE) are shown in Figure 6.3.

6.2.3.1   Explicit Congestion Notification Bits

The first mechanism uses two *explicit congestion notification* (ECN) bits in the frame relay header. They are called the BECN and FECN bits. Figure 6.5 depicts the use of these bits when an FR switch is congested.

Let us suppose that the FR switch B in gray in the figure is approaching a congestion condition. This could be caused by a temporary peak in traffic coming into the node from various sources or a peak in the amount of traffic between B and C. Here is how forward congestion notification would occur:

- FR switch B would detect the onset of congestion based on internal measures such as memory buffer usage or queue length.

- FR switch B would signal FR switch C of the congestion by changing the FECN contained within the frames destined for FR switch C from 0 to 1.

- All intermediate downstream nodes as well as the connected user device (source and destination) would thus learn that congestion is occurring on the DLCIs affected.

It is sometimes more useful to notify the source of the traffic that there is congestion, so that the source can slow down until congestion subsides. This is called backward congestion notification.

In our example, FR switch B looks at frames coming in the other direction on the connection. It sets the BECN bit within those frames to signal the upstream nodes and the attached user device.

6.2.3.2   Discard Eligibility

FR standards state that the user device should reduce its traffic in response to a congestion notification. Implementation of the recommended actions by the user device will result in a decrease in the traffic into the network, thereby reducing congestion. However, if the user device is incapable of responding to the signaling mechanisms, it might simply ignore the congestion signal and continue to transmit data at the same rate as before. This would lead to continued or increased congestion.

When congestion occurs, the network must decide which frames to discard. This is done with the DE bit. When the DE bit is set to 1, it makes the frame eligible for discard in situations of congestion. The DE bit is set to 1 by the frame relay switches when the source transmits frames at a rate exceeding the contracted rate.

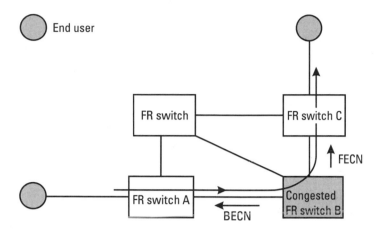

Figure 6.5   Explicit congestion notification mechanism.

6.3   Addressing over Gb

Addressing over Gb is performed over virtual connections. At NS and BSSGP layers, each entity communicates with its peer through this kind of connection. The goal of this section is to explain the hierarchy that has been defined between the different virtual connections and to introduce the technical words that are used in the standard for the description of the Gb interface.

Addressing is very simple, but it requires an understanding of basic concepts that will be introduced in the following sections. Once these concepts have been presented, a global overview of addressing will be given.

Figure 6.6 provides an overview of the different levels of addressing on the Gb interface. We will return to this figure later in the chapter.

6.3.1   Bearer Channel

A *bearer channel* (BC) is a physical channel that carries all the frame relay signaling and data. A BC could be a $n \times 64$ Kbps channel on a pulse code modulation (PCM) link (2,048 Kbps for a European E1 link; 1,544 Kbps for an American T1 link). The PCM link is the typical transmission link that is used in telecom networks.

Note   As a BC can be mapped on a restriction of a PCM link, it is possible for an operator to share this link between GPRS packet transfer (Gb interface) and circuit-switched (A interface).

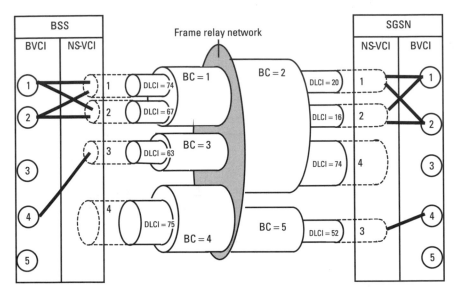

Figure 6.6   Addressing on the Gb interface.

6.3.2   PVC

The frame relay PVC was introduced in Section 6.2. It allows the multiplexing of different flow on the BC. The BC can support several PVCs. At BSS side, a DLCI, which can be different from the one at the SGSN side, identifies a PVC.

Note   There is a dedicated DLCI (DLCI = 0) that is used for signaling purposes (link management). This DLCI does not identify a PVC.

6.3.3   Network Service Virtual Link

An SGSN and a BSS can be connected directly via a physical link or they can be connected indirectly because of intermediate equipment or transmission networks (frame relay network), in which case different physical links are used.

The concept of the *network service virtual link* (NS-VL) is introduced to identify the link defined by a PVC and its supporting BC. Each NS-VL is identified by a *network service virtual link identifier* (NS-VLI). An NS-VL is supported by only one physical link, and several NS-VLs can be mapped on a physical link. One NS-VLI corresponds to the association DLCI and BC identifier.

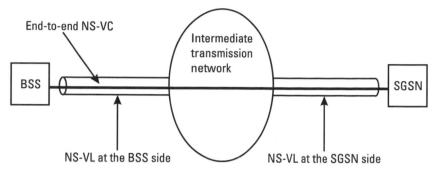

Figure 6.7 Relationship between NS-VCs and NS-VLs. (*From:* [1].)

6.3.4 Network Service Virtual Connection

As described previously, the NS layer has been split into two sublayers, SNS and SNC, in order to have the SNC sublayer independent of the intermediate transmission network (frame relay). In order to provide an end-to-end connection between the BSS and the SGSN irrespective of the exact configuration of the Gb interface and transmission network, the concept of *network service virtual connection* (NS-VC) has been introduced at the SNC layer.

NS-VCs are end-to-end virtual connections between the BSS and the SGSN. Each NS-VC is associated with one PVC at the SNS layer. Each NS-VC is identified by an *NS-VC identifier* (NS-VCI) that has end-to-end significance across the Gb interface. NS-VCs as PVCs are statically configured by the network operator by *operations and maintenance* (O&M) means. Figure 6.7 shows the relationship between NS-VCs and NS-VLs.

6.3.5 Network Service Entity

A *network service entity* (NSE) is composed of a group of NS-VCs; it provides a communication service to the NS user peer entities (SGSN or BSS). A *network service entity identifier* (NSEI) identifies each NSE that has end-to-end significance across the Gb interface. The NSE at each side of the Gb interface manages the traffic from or to a group of cells.

One SGSN can be linked to several BSSs. The NSE and NSEI concept can be used to identify the different BSSs connected to the same SGSN. Each BSS is identified by an NSEI on the SGSN side. A group of NS-VCs is defined within the NSE identified by the NSEI to communicate with the BSS. It could also be possible to define one NSE per board in the BSS or per physical link connected to the BSS. This is completely implementation dependent.

One NSE and the group of NS-VCs that it integrates are defined by the network operator through O&M means. The interest in having several NS-VCs within one NSE is to distribute the traffic from all the cells belonging to this NSE. In case of a problem on one NS-VC, the traffic is not interrupted and is transferred over the other links.

6.3.6   BSSGP Virtual Connection

BSSGP virtual connections (BVCs) are end-to-end virtual connections between the BSS and the SGSN at BSSGP layer. A BVC is identified by a BVCI that has an end-to-end significance across the Gb interface. An NSE is associated with a set of BVCs that are dynamically mapped onto its corresponding NS-VCs.

Two kinds of BVC have been defined:

- PTP BVCs are dedicated to the GPRS traffic of one cell (all the traffic and signaling dedicated to this cell is transmitted via the corresponding BVC);
- Signaling BVCs handle the signaling of the NSE to which they belong.

At the SGSN side, the association BVCI, NSEI identifies one cell.

In the BSS, the BVCIs are statically configured by administrative means. At the SGSN side, BVCIs associated with PTP functional entities are dynamically configured, and BVCIs associated with signaling functional entities are statically configured.

6.3.7   Addressing

Figures 6.6 and 6.8 provide a summary of the different concepts presented in the previous sections. In Figure 6.6, two cells 1 and 2 are identified by BVCI = 1 and BVCI = 2. They share the two NS-VCs identified by NS-VCI = 1 and NS-VCI = 2.

The NS-VC 1 corresponds to the PVC identified by the DLCI = 74 on the BC identified by BCI = 1 at BSS side and DLCI = 20 on the BC identi-

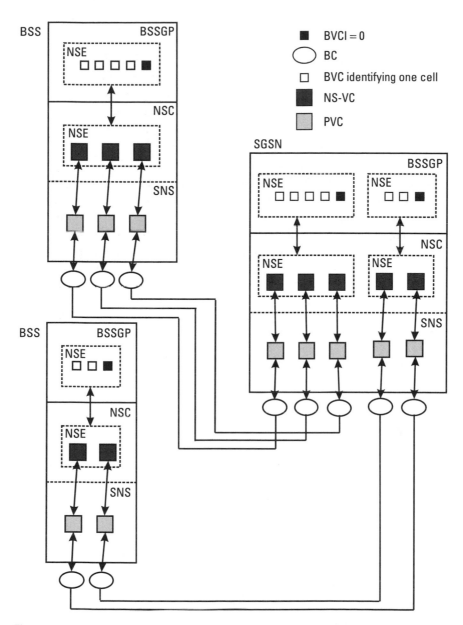

Figure 6.8 Example of addressing between several BSSs and one SGSN.

fied by BCI = 2 at the SGSN side. The NS-VC 2 is mapped on the PVC identified by the DLCI = 67 on the BC identified by BCI = 1 at BSS side and DLCI = 16 on the BC identified by BCI = 2 at SGSN side.

All the traffic addressed to cell 1 or cell 2 from the SGSN or received in cell 1 or cell 2 in the BSS is transferred on the corresponding BVC at the BSSGP layer. The traffic is dynamically distributed on the NS-VCs identified by NS-VCI = 1 and NS-VCI = 2 at the SNC layer.

Figure 6.8 details the addressing between one SGSN and several BSSs. The SGSN is connected to two different BSSs. The BVCs and NS-VCs of each BSS belong to one NSE; each one is identified by a different NSEI.

One BVC is reserved for signaling purposes. On the SGSN side, BVCs and NS-VCs belonging to the same BSS are grouped within the same NSE. One NSE is created for each BSS. For each NSE, a set of BVCs is mapped onto a set of NS-VCs. There is a one-to-one mapping between NS-VC and PVC. Within one SGSN, one cell within one BSS is directly identified by one BVCI and one NSEI.

6.4   NS Layer

6.4.1   SNS Entity

This section describes the SNS sublayer. This sublayer manages the frame relay protocol and consequently the PVCs. The first section explains whether the BSS or the SGSN will behave as the user or network side, depending on the network configuration. The second section describes the signaling procedures that are used at BSS and SGSN sides to signal the addition or deletion of PVCs and to verify the integrity of the link.

6.4.1.1   Network Configuration

The SNS layer is based on frame relay. In a frame relay network, two kinds of interface can be distinguished: the interface between the user and the network (UNI) and the interface between two FR switches of the same network [*network-network interface* (NNI)]. The UNI defines the border between one user and the FR network; the NNI defines the communication between two FR switches.

The Gb interface has been defined as a UNI. However, as two configurations are possible, it is necessary each time to define which node behaves as the user and which node behaves as the network. The two configurations are described in Figure 6.9. The first one corresponds to the case in which there

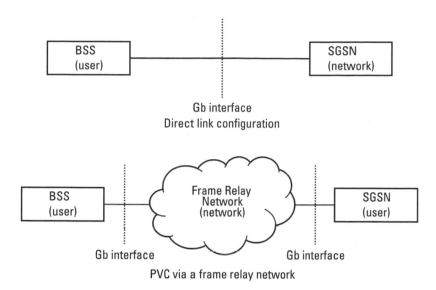

Figure 6.9 Network configurations.

is a direct link between the BSS and the SGSN. In this case, the BSS is considered as the user side of the UNI and the SGSN is considered as the network side.

The second configuration corresponds to the case in which the BSS and the SGSN are connected via an intermediate frame relay network. In this configuration, both BSS and SGSN are treated as the user side of the UNI. The NNI is not directly part of the GPRS network and thus is not defined in the GPRS specifications.

6.4.1.2 Signaling Procedure

This section describes the signaling procedures that are implemented at the SNS layer in order to manage the PVCs between the SGSN and the BSS. The link management involves PVCs status monitoring, detection of newly added PVCs, and link integrity verification. The *link management interface* (LMI) protocol is responsible for these procedures. For GPRS (BSS or SGSN), the protocol is used over the UNI. It is described in [2]. In addition (GPRS standard requirement), it will comply with [3]. The procedures handled by the protocol are as follows:

- Notification of the addition of a PVC;
- Detection of the deletion of a PVC;

- Notification of the availability or unavailability state of a configured PVC;
- Link integrity verification.

These four procedures are handled using the same set of messages. These messages are the STATUS ENQUIRY and STATUS messages. They are sent on DLCI = 0, which identifies signaling.

The STATUS ENQUIRY message is sent to request the status of PVCs or to verify link integrity. It is always sent by the user (the BSS in case of direct link configuration, the BSS and the SGSN otherwise; see previous section). The STATUS message is sent in response to a STATUS ENQUIRY message to indicate the status of PVCs or for link integrity verification.

The status of the PVC connection and the verification of the link integrity are managed using a periodic polling mechanism. This involves periodically requesting, for each user, the link status. Figure 6.10 describes the basic status procedure.

The STATUS ENQUIRY message is composed of:

- The report type indicating the purpose of the message sending. This can be either the request of a full link status or link integrity verification only. When a full link status is requested, the network side indicates the newly added PVCs, the deleted PVCs, and the state of each PVC, and a link integrity verification is performed.
- The link integrity verification parameters. These parameters are described later in this section.

The STATUS message is composed of the report type (indicating either full status or link integrity verification only), the link integrity verification parameters, and the PVCs' status indicating the status of existing PVCs on the BC when it has been requested in the STATUS ENQUIRY message.

Periodic Polling

The user equipment (BSS if there is no intermediate FR network; BSS and SGSN otherwise) always initiates the polling (status enquiry). The polling procedure involves requesting link integrity verification in a periodic way and requesting a full PVC status every N polling cycles (see Figure 6.11). Every N expiries of the timer that controls link integrity verification, the full STATUS ENQUIRY message is sent; the other $N - 1$ expiries trigger

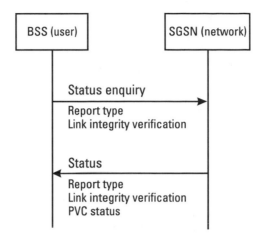

Figure 6.10 Basic status procedure.

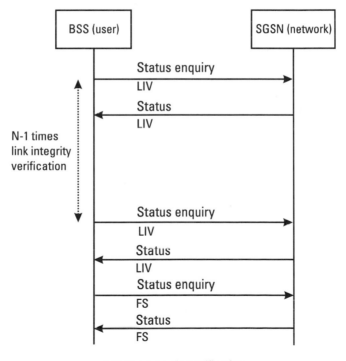

LIV: Link integrity verification
FS: Full status

Figure 6.11 Periodic polling procedure.

the sending of the "link integrity verification only" STATUS ENQUIRY message.

In response to a full status request, the network sends a STATUS message containing a PVC status information element for each PVC configured on that BC. Each information element contains an active bit indicating the availability or unavailability of that PVC. When a PVC is detected as nonactive, the user equipment stops transmitting frames on the PVC until the PVC becomes active again. When the user receives a full status message containing a PVC status information element identifying an unknown DLCI and indicating a new PVC, the user equipment marks this PVC as new.

Link Integrity Verification

The purpose of the link integrity verification is to allow both sides of the UNI to determine the status of the in-channel signaling link (DLCI 0). This is necessary, since these procedures use *unnumbered information* (UI) frames; these frames are not acknowledged between the peer entities.

For the link integrity verification procedure, each side of the UNI maintains one *receive sequence counter* (RSC) and one *send sequence counter* (SSC). Every time a STATUS ENQUIRY or STATUS message is sent, the values of these counters are included in the message by the sending side. The field *send sequence number* (SSN) within the message indicates the value of the SSC, and the *receive sequence number* (RSN) indicates the value of the RSC.

The SSC indicates the number of STATUS ENQUIRY or STATUS messages that have been sent. The RSC indicates the number of STATUS ENQUIRY or STATUS messages that have been received.

Before sending a STATUS ENQUIRY message, the user side increments its SSC counter. When the network side receives the STATUS ENQUIRY message, it compares the RSN received within the message with its own SSC. The number of messages that have been received at one side will be equal to the number of messages that have been sent from the other side.

The received SSN is stored in the RSC. The network side increments its SSC counter and sends the STATUS message. At the reception of the STATUS message, the user side compares the RSN received with its own SSC. Whenever the SSC and the RSN values do not match, a message has been lost on the interface and the integrity of the link is not verified.

Figure 6.12 gives an example of the link integrity verification procedure. When the BSS receives the STATUS message, it compares the number of STATUS ENQUIRY messages that have been received at the SGSN side (RSN = 1) with the number of STATUS ENQUIRY messages that it has

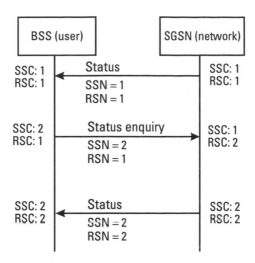

Figure 6.12 Example of successful link integrity verification procedure.

sent (SSC = 1). The number of STATUS messages received is then equal to the number of STATUS messages sent by the SGSN (RSC is set to SSN).

Before sending a new STATUS ENQUIRY message, the BSS increments its SSC counter. The comparison of the number of STATUS ENQUIRY messages received and sent by the BSS is then performed at SGSN side.

6.4.2 NSC Entity

The NSC sublayer is responsible for the management of the NS-VCs between the BSS and the SGSN and the transfer of upper-layer packets. This section details the procedures that are used at NSC layer for the transfer of upper-layer data and the procedures that are used for the management of the NS-VCs.

6.4.2.1 Upper-Layer Packet Transfer

The NS-UNITDATA message is used to transfer upper-layer packets or SDUs that are encapsulated inside. This message is sent across the Gb interface in unacknowledged mode. The same message is used for transfer from the BSS to the SGSN or from the SGSN to the BSS (see Figure 6.13). The NS-UNITDATA message contains the SDU to be transmitted and the BVCI identifying the addressed BVC.

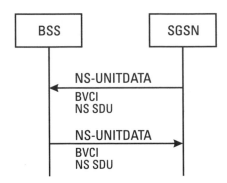

Figure 6.13 NS-PDUs transfer.

6.4.2.2   NS-VC Management

The NSC sublayer is responsible for the management of NS-VCs. These are statically configured by O&M means.

One NS-VC can be in one of the following three states:

- *DEAD*. In this state end-to-end communication between the peer NS entities does not exist.
- *BLOCKED* & *ALIVE*. In this state the NS-UNITDATA are not allowed to transit across the NS-VC. This could be due to an O&M intervention or equipment failure.
- *UNBLOCKED*. In this state NS-UNITDATA are accepted and can be sent.

Figure 6.14 describes the NS-VC state transition. After a successful reset procedure, the NS-VC state is BLOCKED & ALIVE. The transition from the BLOCKED state to the UNBLOCKED state and the reverse transition are managed by the unblocking and blocking procedures, respectively. As soon as an NS-VC has been reset (the NS-VC is in BLOCKED or UNBLOCKED state), a periodic test procedure is triggered to regularly verify its availability.

NS-VC Reset

This procedure is used when a new NS-VC is set up, after a processor restart, a failure recovery, or any local event restoring an existing NS-VC in DEAD or undetermined state.

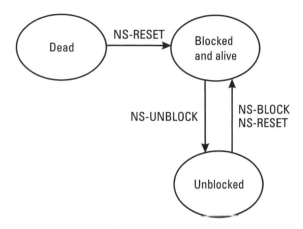

Figure 6.14 NS-VC state transition within the BSS and SGSN.

At the end of the procedure, a successful reset NS-VC is marked as BLOCKED and ALIVE. The initiator of the procedure must trigger the periodic test procedure.

This procedure may be initiated by the SGSN or by the BSS. Figure 6.15 shows an example of NS-VC reset initiated by the BSS. The BSS sends an NS-RESET message to the SGSN. This message indicates the cause of the reset procedure, the NSEI, and NS-VCI identifying the NS-VC to be reset. The NS-VC is then marked as BLOCKED and DEAD within the BSS.

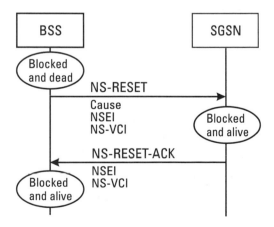

Figure 6.15 Reset procedure.

When the SGSN receives the NS-RESET message and if it is able to reset the NS-VC, it returns an NS-RESET-ACK message indicating the NS-VCI and NSEI identifying the NS-VC that has been reset. This message is sent on the NS-VC that has been reset. The NS-VC is marked as BLOCKED and ALIVE at SGSN side. When the BSS receives the NS-RESET-ACK message, it marks the NS-VC as BLOCKED and ALIVE and it initiates the test procedure. After the reset of an NS-VC, the unblocking procedure of the NS-VC is triggered by the originator of the reset procedure.

Blocking/Unblocking of an NS-VC

The blocking/unblocking procedures are used in case of failure in the transmission network, O&M intervention, or equipment failure. These procedures can be triggered either by the SGSN or the BSS.

Blocking Procedure   Figure 6.16 shows an NS-VC blocking procedure triggered by the SGSN. Once the decision to block an NS-VC has been made, the SGSN (in the example) marks the NS-VC as BLOCKED. The NS-UNITDATA messages will be redistributed to other UNBLOCKED NS-VCs belonging to the same NSE, thanks to the load-sharing function (see Section 6.4.2.3).

The NS-BLOCK message is sent by the SGSN. It contains the NS-VCI of the NS-VC to be blocked. The NS-BLOCK message can be sent in any alive NS-VC pertaining to the NSE. The SGSN will continue to accept

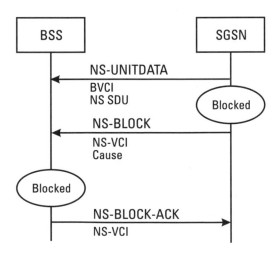

Figure 6.16 Example of blocking procedure initiated by the SGSN.

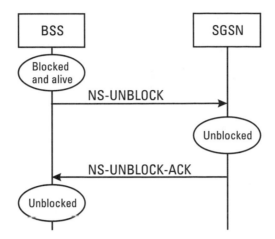

Figure 6.17 Example of unblocking procedure between the SGSN and the BSS.

NS-UNIDATA messages until it receives the NS-BLOCK-ACK PDU indicating the NS-VC blocking.

On the BSS side, at the reception of the NS-BLOCK message, the NS-VC is marked as BLOCKED in the BSS. The NS user entity is informed and NS-UNITDATA messages are redistributed to other UNBLOCKED NS-VCs. The NS-BLOCK-ACK message is sent in any ALIVE NS-VC pertaining to the same NSE.

Unblocking Procedure   Figure 6.17 shows an NS-VC unblocking scenario triggered by the BSS. The BSS triggers the unblocking of an NS-VC by sending an NS-UNBLOCK message on the concerned NS-VC. For this the NS-VC will be in the ALIVE state (see the following section). When the SGSN receives the NS-UNBLOCK message and if it is able to unblock the NS-VC, it returns an NS-UNBLOCK-ACK message on the same NS-VC. The NS-VC state is changed to UNBLOCKED. The load-sharing function and the NS user entity are informed. In the BSS, the state of the NS-VC is changed at the reception of the NS-UNBLOCK-ACK PDU.

NS-VC Test

The test procedure is used when one NSC entity wishes to check that end-to-end communication with its peer entity exists on an NS-VC. This procedure is triggered at the end of a successful reset procedure by the originator of the reset procedure and is periodically repeated. The procedure can be ini-

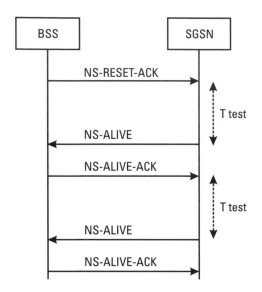

Figure 6.18 Test procedure initiated by the SGSN.

tiated by the BSS or the SGSN, depending on the originator of the reset procedure.

Figure 6.18 illustrates a test procedure initiated by the SGSN. On
receipt of the NS-RESET-ACK message, the originator (SGSN) of the reset
procedure starts a timer (Ttest) managing the periodical repetition of the test
procedure. When this timer expires, the originator sends an NS-ALIVE message on the NS-VC to be checked. Upon receipt of the NS-ALIVE message
on an ALIVE NS-VC, the BSS returns an NS-ALIVE-ACK message on the
same NS-VC. Upon receipt of the NS-ALIVE-ACK message, the SGSN
restarts the timer managing the repetition of the test procedure.

Note that when a failure in the test procedure is detected, it is restarted.
After several failures of the test procedure, the NS-VC is marked as
BLOCKED and a blocking procedure is triggered.

6.4.2.3 Load Sharing

The load-sharing function of an NSE receives as input the NS SDUs from all
the BVCs belonging to this NSE (see Figure 6.19). It is responsible for the
NS SDU traffic repartition between the NS-VCs belonging to this NSE. It
provides to the upper layer a seamless service upon failure. In fact, in such a
case, it reorganizes the NS SDU traffic between the UNBLOCKED NS-VCs
of the same NSE. This procedure is implemented at both the BSS and

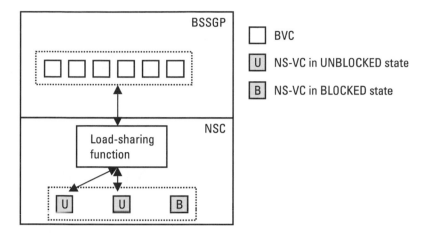

Figure 6.19 Load-sharing function.

SGSN sides. Upon reorganization of the traffic, the order of NS SDU transfer may be disturbed.

All NS SDUs to be transmitted over Gb are passed to the load-sharing function along with the *link selector parameter* (LSP), the BVCI, and the NSEI. The LSP and BVCI are used to select the UNBLOCKED NS-VCs within the group. For each BVCI and NSEI, the selection of the NS-VC is based on the LSP. For each BVCI and NSEI, NS SDUs with the same LSP will be sent on the same NS-VC. The load-sharing function guaranties that for each BVC, the order of all NS SDUs marked with the same LSP value is preserved.

Note that on the Gb interface, upper-layer SDUs for the same mobile must be transferred in order. The LSP is a parameter, implementation dependent, that is used to guaranty this. In the SGSN, as each mobile is uniquely identified by its TLLI, it could be possible to use the TLLI as the LSP parameter. This will guaranty ordered transfer for each mobile.

6.5   BSSGP Layer

This section describes the BSSGP layer. It is responsible for the management of the BVC that corresponds to the virtual connection in which all the signaling and data addressed to a particular cell is sent through.

The first section explains the layers to which BSSGP provides services. In the following sections, the procedures that are used to provide services to these layers are detailed. The second section describes the procedures that are used on the interface for the transfer of user data in both uplink and downlink. The third section describes the procedures that are used for the transfer of GMM signaling. The fourth section handles the procedure used for the management of the BVCs. The last section deals with PFM procedures.

6.5.1   BSSGP Service Model

Figure 6.20 describes the BSSGP service model. It presents all the layers to which BSSGP provides services, together with their *service access points* (SAPs). It provides services to the following layers:

- *LLC*. BSSGP provides functions that control the transfer of LLC frames passed between the SGSN and the MS across the Gb interface.
- *GMM*. BSSGP provides functions associated with mobility management between an SGSN and a BSS.
- *NM*. This concerns the functions associated with the management of the virtual connections between the SGSN and BSS and the control of the BSS or SGSN nodes such as flow control.
- *Relay (RL)*. This layer acts as relay between the Gb interface and the radio interface. It controls the transfer of LLC frames between the RLC/MAC and BSSGP layers.
- *PFM*. This handles the management of BSS packet flow contexts between the SGSN and the BSS.

6.5.2   User Procedures Between LLC and RELAY SAPs

Figure 6.21 describes the procedures used for the transfer of LLC PDUs between the SGSN and the BSS.

6.5.2.1   Transfer of LLC PDUs in Downlink

The transfer of LLC PDUs in downlink is performed in unacknowledged mode. The LLC PDU is transported within the DL-UNITDATA message. This message is sent on the BVC of the cell in which is located the addressed mobile.

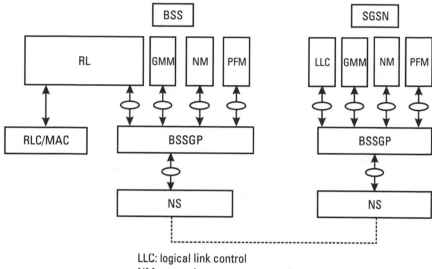

LLC: logical link control
NM: network management
PFM: packet flow management
RL: relay
GMM: GPRS mobility management
⬭ SAP: service access point

Figure 6.20 BSSGP service model.

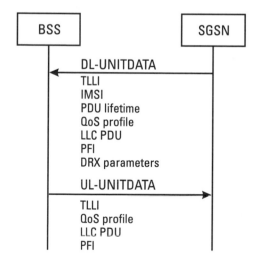

Figure 6.21 Data PDUs transfer.

The SGSN provides the BSS with MS-specific information, enabling the RLC/MAC entity in the BSS to transmit the LLC PDU to the MS in a user-specific manner. The information given to the RLC/MAC entity include:

- *MS radio access capability.* This defines the radio capability of the mobile (multislot class, power capability, frequency band supported).

- *PFI.* This parameter identifies the packet flow context (when existing), containing the QoS parameters for the transfer of the LLC PDU.

- *QoS profile.* This defines the peak bit rate, the type of BSSGP SDU (signaling or data), the type of LLC frame, the precedence class, and the transmission mode (ACK NACK, see Chapter 5).

- *PDU lifetime.* This defines the remaining time period that the PDU is considered as valid within the BSS. If the PDU is held for a period exceeding the PDU lifetime time period, the PDU is discarded by the BSS.

- *DRX parameters.* These parameters are used by the BSS to determine when the mobile operates in DRX or non-DRX mode.

The DL-UNITDATA message contains the current TLLI identifying the MS, the IMSI, the QoS profile, the PDU lifetime, and optionally the PFI, the MS radio access capability, and the DRX parameters. If the SGSN has valid DRX parameters (see Chapter 3) for a TLLI, it includes them in the PDU.

Note that during GMM attach procedure, it may happen that the SGSN does not have a valid IMSI, in which case the DL-UNITDATA PDU is sent without the IMSI.

The TLLI allows the BSS to identify the DL transfer. It could happen that during an uplink transfer on the air interface, the BSS receives a downlink LLC PDU that is addressed to the mobile performing the uplink transfer. The TLLI is used by the BSS to correlate the uplink and downlink transfer so that it allocates downlink resources that when taken in combination with the allocated uplink resources remain within the multislot class constraint of the mobile.

Note that the NSEI and the BVCI are not contained in the DL-UNIT-DATA message. The NS layer provides them together with the DL-UNIT-DATA message when it is received.

6.5.2.2 Transfer of LLC PDUs in Uplink

The transfer of LLC PDUs in uplink is performed in unacknowledged mode. One LLC PDU is encapsulated in one UL-UNIDATA PDU. This PDU is sent on the BVC of the cell in which it has been received.

The UL-UNITDATA message contains:

- The TLLI identifying the mobile that has sent the LLC PDU;
- The QoS profile that has been used for the transfer of the LLC PDU on the air interface;
- The LLC PDU;
- The PFI identifying the packet flow context associated with the transfer (optional).

Note that the NSEI and BVCI are not part of the UL-UNIDATA message. The NS layer provides them together with the DL-UNITDATA message when it is received.

6.5.3 Signaling Procedures Between GMM SAPs

6.5.3.1 Paging

Two kinds of paging procedures can be initiated by the SGSN:

1. Paging procedure for GPRS services, in which case the SGSN can send one or more PAGING-PS PDU to the BSS;
2. Paging procedure for non-GPRS services requested by the MSC/ VLR, in which case the SGSN can send one or more PAGING-CS PDUs to the BSS.

These paging PDUs contain information that is necessary for the BSS to initiate the paging for an MS within a group of cells. The level of resolution of this group of cells can be:

- All cells within the BSS;
- All cells within one LA;
- All cells within one RA;
- One cell (one BVCI).

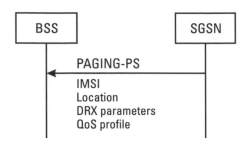

Figure 6.22 Paging PS procedure.

When the mobile is in GMM READY state, the cell in which the mobile is located is provided in the paging message. The paging message is then sent on the BVC associated with the cell.

When the mobile is in GMM IDLE state, the paging message contains one of the following locations for the mobile: LA, RA, or BSS. If the LA, RA, or BSS in which the mobile is located is mapped on different NS entities within the SGSN, it must send one paging message to each NS entity having cells that belong to the LA, RA, or BSS. In this case, the paging message is sent on each signaling BVC of the NSEs.

Figure 6.22 describes a paging PS procedure. The PAGING-PS PDU contains the following parameters:

- The IMSI of the paged mobile. It is used by the BSS to derive the PAGING GROUP of the mobile.
- The location of the mobile: either cell, RA, LA or BSS.
- The QoS profile.
- The DRX parameters when available at SGSN side. They inform the BSS of the non-DRX and DRX period of the mobile.
- The PFI when available. It indicates the associated packet flow context.
- The P-TMSI when available at the SGSN side. When it is present in the message, it is used to page the mobile on the radio interface. Otherwise, the IMSI is used.

In the case of paging for non-GPRS services, the SGSN provides the IMSI of the mobile, its location, and the DRX parameters. The TLLI and TMSI can also be provided by the SGSN. In this case, one of the parameters

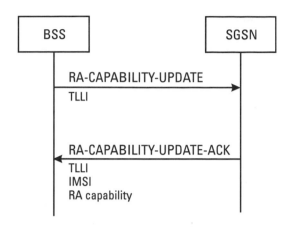

Figure 6.23 Radio access capability update procedure.

is used to page the mobile on the radio interface. The IMSI and DRX parameters are used by the BSS to derive the PAGING GROUP.

A PAGING-PS or PAGING-CS PDU is relayed in the BSS by a PACKET PAGING REQUEST message when PCCCH is present in the cell; otherwise by a PAGING REQUEST message that is sent on the air interface.

6.5.3.2 Radio Access Capability Update

The BSS triggers the RA capability update procedure to request an MS's current RA capability, its IMSI, or both. This is requested by sending an RA-CAPABILITY-UPDATE PDU (see Figure 6.23). This message includes the TLLI of the MS. It is sent on the BVC associated with the cell where the mobile is located.

The SGSN responds by sending an RA-CAPABILITY-UPDATE-ACK including the TLLI, the IMSI, and the RA capability if available.

6.5.3.3 Radio Status

This procedure is triggered by the BSS when it has been requested to send downlink LLC PDUs and, because of exception conditions, the transfer through the air interface of LLC PDUs is no longer possible. Exception conditions are occurrences in which:

1. The MS goes out of coverage and is lost;

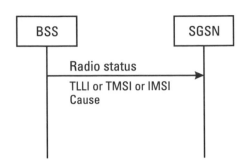

Figure 6.24 Radio STATUS procedure.

2. The link quality is too poor to continue the transfer through the air interface;

3. The BSS has ordered the MS to perform a cell reselection.

Conditions 1 and 2 indicate to the SGSN that it should stop sending LLC PDUs to the cell for the concerned MS. Condition 3 indicates to the SGSN that it should wait for a cell update before resuming the transmission of LLC PDUs.

As described in Figure 6.24, the BSS uses the RADIO-STATUS PDU to signal these exception conditions to the SGSN. The RADIO STATUS PDU includes the identification of the MS (either the TLLI, IMSI, or TMSI) and the cause of the failure. It is sent on the BVC associated with the cell where the mobile was located.

6.5.4   Signaling Procedures Between NM SAPs

6.5.4.1   Flush

The flush procedure is initiated by the SGSN. This procedure is triggered when the SGSN detects that a mobile has performed a cell reselection during a downlink packet transfer. The SGSN detects the cell reselection after a cell update or RA update procedure. The procedure can be initiated for two purposes:

1. The cell reselection occurs between two cells that are not in the same RA or in the same NS entity, in which case the flush procedure ensures that LLC PDUs that are stored in the BSS at cell level, for the concerned mobile, are deleted.

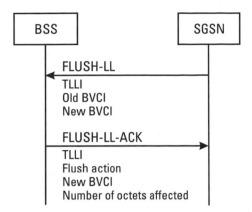

Figure 6.25 Flush procedure.

2. The cell reselection occurs between two cells that are in the same RA and in the same NS entity, in which case the flush procedure indicates the new cell (BVCI) where the LLC PDUs are to be transferred. The interruption of traffic due to the cell reselection is then reduced.

The flush procedure is described in Figure 6.25. The SGSN initiates the flush procedure by sending a FLUSH-LL PDU to the BSS on the NSE signaling BVC. This message contains the flowing parameters:

- The TLLI identifying the transfer;
- The BVCI (old) for which the LLC PDUs must be deleted;
- The BVCI (new) in which the LLC PDUs must be transferred (the new BVCI is only indicated when the two BVCI belongs to the same RA and NSE).

This procedure is acknowledged by the FLUSH-LL-ACK message, which contains the following parameters:

- The TLLI;
- The flush action indicating whether the LLC PDUs have been deleted or transferred toward the new BVCI;
- The new BVCI in case of transfer of the LLC PDUs;

- The number of octets that have been deleted or transferred (this parameter is used to calibrate the flow control algorithm that will be described in Section 6.5.4.3).

6.5.4.2 LLC Discarded

The LLC discarded procedure is described in Figure 6.26. In the procedure, the BSS sends an LLC-DISCARDED PDU on the NSE signaling BVC in order to inform the SGSN that an LLC PDU has been deleted within the BSS for one BVCI. This may occur, for instance, at a PDU lifetime expiry or cell reselection. The LLC-DISCARDED PDU includes the following parameters:

- The TLLI identifying the mobile;
- The BVCI that is affected;
- The number of LLC frames that have been discarded;
- The number of octets that have been deleted.

These two last parameters are used for setting the flow control algorithm that is used between the SGSN and the BSS. This mechanism is described in the next section.

6.5.4.3 Flow Control

A flow control procedure between the BSS and the SGSN is necessary in order to manage buffers within the BSS. In the BSS, there is at least one buffer for each BVC and possibly for each MS. In order to avoid DL LLC PDU loss because of buffer overflow, the BSS controls the transfer of BSSGP UNITDATA PDUs for an MS.

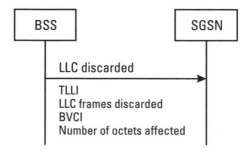

Figure 6.26 LLC discarded procedure.

Only downlink BSSGP UNITDATA PDU transfer is managed via a flow control procedure. There is no uplink flow control. Buffers and link capacity must be dimensioned in order to avoid loss of uplink data.

The basic principle of BSSGP flow control is that the BSS sends to the SGSN flow control parameters that allow the SGSN to locally control its transmission in the BSS direction. The BSS controls the flow of BSSGP UNIDATA PDUs by indicating the maximum allowed throughput in total for each BVC and for each MS.

The SGSN passes LLC PDUs to the BSS as long as the allowed BSSGP throughput is not exceeded. The allowed BSSGP throughput is given per BVCI and for a single MS on that BVCI. The SGSN schedules the BSSGP downlink traffic of all MSs of a BVCI according to the maximum and guaranteed bit rate attributes and to the QoS profile related to each LLC PDU.

Flow Control Scenario

The SGSN performs flow control at two levels: BVC level and MS level. The flow control is performed on each DL LLC PDU first at MS flow control level then at BVC flow control level.

If a DL LLC PDU is allowed to pass by an individual MS flow control, the SGSN then applies the BVC flow control to the DL LLC PDU. If a DL LLC PDU is passed by both flow control mechanisms (MS level and BVC level), the DL LLC PDU is delivered to the NS layer for transmission to the BSS.

The BSS evaluates flow control parameters for each MS (respectively each BVC) and sends them to the SGSN in the FLOW-CONTROL-MS (respectively, FLOW-CONTROL-BVC) PDUs. They are sent on the BVC associated with the cell. These flow control parameters provided by the BSS are used by the SGSN to tune and update its flow control algorithm.

Figure 6.27 describes the overall process of flow control at BSSGP layer. At the SGSN side, there is a hierarchical flow control mechanism that is applied first at MS level and cell level (BVC). The MS flow control mechanism is controlled by the BSS by using the message FLOW-CONTROL-MS. The BVC flow control is managed by the BSS by sending FLOW-CONTROL-BVC messages. At BSS side, there is a BVC flow control context that contains all the flow control contexts of the MSs that are located in the cell corresponding to the BVC.

Upon receipt of FLOW-CONTROL-MS (respectively, FLOW-CONTROL-BVC) PDU, the SGSN sends a FLOW-CONTROL-MS-ACK (respectively, FLOW-CONTROL-BVC-ACK). A timer whose value is com-

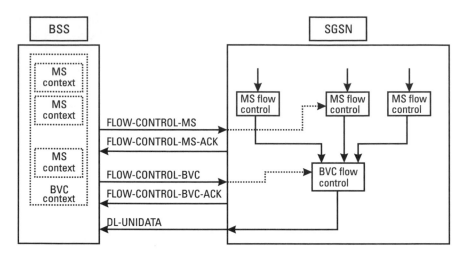

Figure 6.27   BSSGP flow control procedure.

mon to the BSS and the SGSN limits the rate at which the BSS is allowed to send flow control messages.

LLC PDUs queued within the BSS that are not transferred across the radio interface because of the PDU lifetime expiration are deleted. The SGSN is notified by an LLC-DISCARDED PDU that allows the BVC and MS flow control context to be updated.

Flow Control Algorithm

The BSSGP flow control mechanism is based on a bucket algorithm that is described below.

In principle, the BSS indicates the amount of memory available for each flow context within the BSS (for each BVC and each mobile within this BVC). This amount of memory is given by the maximum bucket size parameter (B_{max}).

Every time an LLC PDU must be sent in downlink toward the BSS, the flow control mechanism at SGSN side verifies that if this PDU is sent to the BSS the maximum bucket size for the mobile and BVC context will not be reached. If both maximum bucket sizes are not reached, the PDU is sent toward the BSS.

However, the bucket on the BSS side empties as LLC PDUs are sent on the radio interface. In order to be able to estimate the bucket size at any time, the SGSN needs some information about the bucket leak rate (R) at mobile and BVC levels.

The flow control parameters that are provided by the BSS within the flow control PDUs are

- The bucket size (B_{max});
- The bucket leak rate (R);
- The bucket leak ratio (this parameter has been introduced as an optional feature and will be supported by both sides of the Gb in order to be used.

The two first parameters are used to tune the algorithm in the SGSN and the last one is used to realign the bucket counter. It gives the exact amount of available space within the bucket.

Note that the FLOW-CONTROL-MS message also contains the TLLI identifying the mobile context. The FLOW-CONTROL-BVC message contains the default MS bucket parameters that must be used by the SGSN before it receives the first FLOW-CONTROL-MS message for the mobile.

Every time a new LLC PDU arrives, the flow control algorithm estimates the new value of the bucket counter (B^*) if the LLC is passed.

$$B^* = B + \text{Length}(\text{LLC PDU}) - R(\text{Tarrival} - \text{Tlast\_llc\_transferred}) \qquad (6.1)$$

where

B stands for the evaluated bucket size at the time the last LLC PDU was transferred (Tlast_llc_transferred).

Tarrival is the time of arrival of the LLC PDU for which the flow control is triggered.

R (Tarrival – Tlast_llc_transferred) represents the bucket leak amount between the reception of the new LLC PDU and the transmission of the last one.

If the new value of the bucket counter (B^*) is lower than B_{max}, the LLC PDU is passed; otherwise, it is delayed.

The value of the bucket counter must be updated upon receipt of a FLUSH-LL-ACK PDU or LLC-DISCARDED PDU indicating that LLC PDUs have been deleted for one cell or have been transferred toward a new BVCI.

6.5.4.4   BVC Management

A BVC is a virtual connection that handles the traffic of one cell between the SGSN and the BSS. It could be in one of the following two states: BLOCKED state or UNBLOCKED state. When the BVC is BLOCKED, there is no activity in the cell.

The BSS is responsible for the management of the BVC state. This means that it always initiates blocking or unblocking procedures that bring about the change of state of the BVC. One BVC could be blocked because of O&M intervention in a cell or equipment failure at the BSS side.

Figure 6.28 shows the BVC state transition diagram in the BSS and the SGSN. At BSS side, the transition between the two states is always triggered by O&M intervention. Within the SGSN, the transition between the two states is triggered by the blocking/unblocking procedure, which is always triggered by the BSS or the reset procedure.

BVC Blocking and Unblocking Procedure

The blocking of a BVC is initiated by the BSS by sending a BVC-BLOCK PDU on the NSE signaling BVC. This message contains the BVCI identifying the BVC that must be blocked and the cause of the blocking. On receipt of the BVC-BLOCK PDU, the SGSN marks the indicated BVC as BLOCKED and stops transmitting traffic addressed to this BVC. It acknowledges the blocking by sending a BVC-BLOCK-ACK PDU that contains the BVCI of the BLOCKED BVC on the NSE signaling BVC.

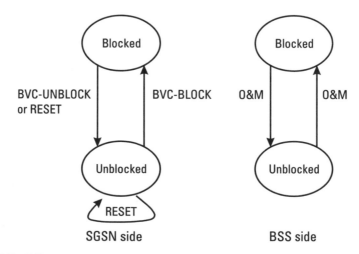

Figure 6.28   BVC state transition in the BSS and SGSN.

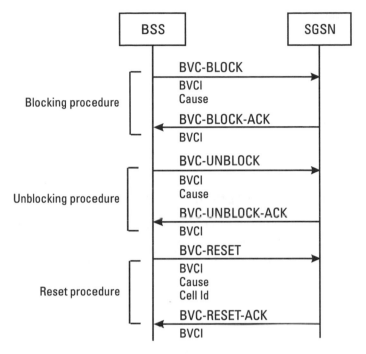

Figure 6.29 BVC blocking/unblocking and reset procedures.

The BSS initiates the unblocking of the BVC by sending a BVC-UNBLOCK PDU containing the BVCI that identifies the BVC to be UNBLOCKED and the cause of the BVC unblocking. This message is sent on the NSE signaling BVC. Upon receipt of this message, the SGSN marks the BVC as UNBLOCKED and sends a BVC-UNBLOCK-ACK message containing the BVCI of the UNBLOCKED BVC. These procedures are described in Figure 6.29.

Note that the blocking/unblocking procedure is not applicable to the signaling BVC. It will never be in the BLOCKED state.

BVC Reset Procedure

The reset procedure can be initiated either by the SGSN or the BSS. It is used to synchronize the initialization of GPRS BVC-related contexts at the BSS or SGSN. It is used by the BSS to map one BVCI on one cell.

It could be performed because of recovery procedures related to a system failure in the SGSN or BSS, an underlying NS system failure, a change in mapping between BVCI and CI, and creation of a new BVC. After a BVC reset procedure, the affected BVC is assumed to be UNBLOCKED at the SGSN side.

When one side sends a BVC-RESET PDU it will stop sending PDU until it receives the acknowledgment. The other side initializes all BVCs belonging to the NSE and returns a BVC-RESET-ACK PDU on the NSE signaling BVC. The BVC-RESET PDU contains the BVCI identifying the BVC, the corresponding CI, and the cause of the reset procedure. The BVC-RESET-ACK message contains the BVCI identifying the reset BVC. The reset procedure is described in Figure 6.29.

6.5.5   Signaling Procedures Between PFM SAPs

PFM procedures are used to create, modify, or delete a packet flow context within the BSS. These procedures are triggered every time a PDP context procedure is invoked at SGSN side. The definition of the packet flow context is in Chapter 3.

A TFI identifies each packet flow context, and the validity of the BSS packet flow context is managed by a timer within the BSS (packet flow timer PFT).

The BSS packet flow context procedures are as follows:

- BSS packet flow context creation procedure;
- BSS packet flow context modification procedure;
- BSS packet flow context deletion procedure.

6.5.5.1   BSS Packet Flow Context Creation

The BSS packet flow context creation procedure is used to create a BSS packet flow context within the BSS. It can be initiated either by the SGSN or by the BSS.

The SGSN may request the creation of a BSS packet flow context at the activation of a PDP context.

The BSS packet flow context is negotiated by the BSS with the SGSN at the transmission of an uplink or downlink LLC PDU associated with an unknown PFI.

The procedure is described in Figure 6.30. The DOWNLOAD-BSS-PFC is sent only when the procedure is initiated by the BSS.

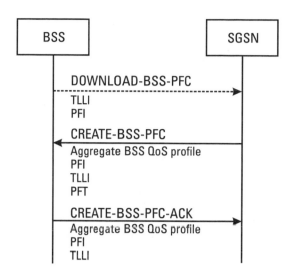

Figure 6.30 BSS packet flow context creation.

Note that when the BSS is unable to create the PFC, it returns a CRE-ATE-BSS-PFC-NACK PDU that contains the TLLI, the PFI, and the cause.

6.5.5.2 BSS Packet Flow Context Modification

This procedure can be initiated by the SGSN or the BSS. When initiated by the SGSN, the procedure is the same as for BSS packet flow context creation. It can be triggered due to an increase or decrease of resources at the BSS side. Figure 6.31 describes the procedure when initiated by the BSS.

6.5.5.3 BSS Packet Flow Context Deletion

This procedure is initiated by the SGSN. It may happen that the BSS deletes a packet flow context (e.g., due to a memory problem). However, in this case no notification issent to the SGSN. The procedure is described in Figure 6.31.

6.5.6 Summary of BSSGP PDUs

Table 6.1 gives for each BSSGP PDU its usage, the BVC type on which it is sent, and the direction.

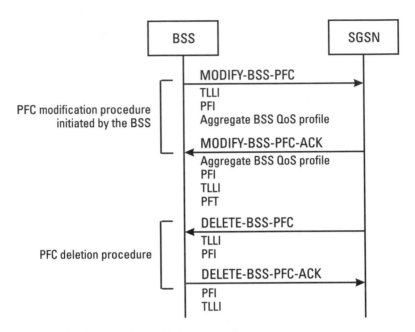

Figure 6.31 BSS PFC modification and deletion procedures.

Table 6.1
Summary of BSSGP PDUs

PDU Name	Usage	BVC Type	Direction*
DL-UNITDATA	LLC	Cell BVC	DL
UL-UNITDATA	LLC	Cell BVC	UL
PAGING-PS	GMM	Signaling BVC or Cell BVC	DL
PAGING-CS	GMM	Signaling BVC or Cell BVC	DL
RA-CAPABILITY-UPDATE	GMM	Cell BVC	UL
RA-CAPABILITY-UPDATE-ACK	GMM	Cell BVC	DL
RADIO-STATUS	GMM	Cell BVC	UL
FLUSH-LL	NM	Signaling BVC	DL
FLUSH-LL-ACK	NM	Signaling BVC	UL
LLC-DISCARDED	NM	Signaling BVC	UL
FLOW-CONTROL-MS	NM	Cell BVC	UL

*DL refers to the direction from SGSN to BSS, UL refers to the reverse direction.

Table 6.1 (continued)

PDU Name	Usage	BVC Type	Direction*
FLOW-CONTROL-MS-ACK	NM	Cell BVC	DL
FLOW-CONTROL-BVC	NM	Cell BVC	UL
FLOW-CONTROL-BVC-ACK	NM	Cell BVC	DL
BVC-BLOCK	NM	Signaling BVC	UL
BVC-BLOCK-ACK	NM	Signaling BVC	DL
BVC-UNBLOCK	NM	Signaling BVC	UL
BVC-UNBLOCK-ACK	NM	Signaling BVC	DL
BVC-RESET	NM	Signaling BVC	UL or DL
BVC-RESET-ACK	NM	Signaling BVC	DL or UL
CREATE-BSS-PFC	PFM	Signaling BVC	DL
CREATE-BSS-PFC-ACK	PFM	Signaling BVC	UL
CREATE-BSS-PFC-NACK	PFM	Signaling BVC	UL
MODIFY-BSS-PFC	PFM	Signaling BVC	UL
MODIFY-BSS-PFC-ACK	PFM	Signaling BVC	DL
DELETE-BSS-PFC	PFM	Signaling BVC	DL
DELETE-BSS-PFC-ACK	PFM	Signaling BVC	UL
DOWNLOAD-BSS-PFC	PFM	Signaling BVC	UL

*DL refers to the direction from SGSN to BSS, UL refers to the reverse direction.

6.6   Case Studies

6.6.1   Establishment of a BVC

This section describes the procedure for a BVC establishment between the SGSN and the BSS. The procedure is described in Figure 6.32. It is triggered by O&M. For each cell in the BSS supporting GPRS service, a BVC must be established between the BSS and the SGSN in order to exchange GPRS traffic.

The establishment procedure is only possible if at least one NS-VC at NS layer, belonging to the same NSE of the cell, already exists and is operational. The BSS starts the establishment of the BVC by triggering a BVC reset procedure. At the end of the procedure the BVC is in the UNBLOCKED state.

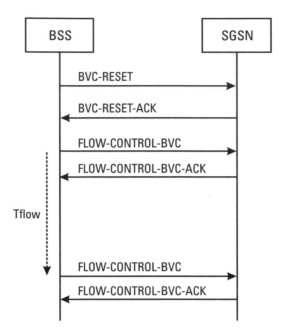

Figure 6.32 BVC establishment procedure.

Once the BVC has been established, the BSS must indicate the amount of memory available for the BVC for flow control between the SGSN and the BSS. This will allow downlink traffic on the BVC. Thus the BSS triggers a BVC flow control procedure indicating bucket parameters necessary for initiating the flow control procedure. The basic flow control procedure is repeated in a periodic way (controlled by the T_{flow} timer in the figure).

Note that as described in Section 6.5.4.4, at the end of a successful BVC reset procedure, the BVC is in the UNBLOCKED state. If the BSS does not want to allow the traffic immediately after the reset procedure, therefore not starting the flow control procedure, it must block the BVC.

Every time the BSS sends a FLOW-CONTROL-BVC message to the SGSN, it must indicate the bucket size B_{max} of the BVC and the bucket leak rate R. B_{max} must be evaluated so that no underflow condition occurs during downlink transfer because it is undersized. The bucket size B_{max} is dependant of the following cell parameters:

- The number of time slots in the cell that carries GPRS traffic (PDCH);

- The maximum throughput that can be achieved per PDCH.

B_{max} can be computed as follow:

$$B_{max} = MAXth * Npdch * Tflow (1) \qquad (6.2)$$

where

> *MAXth* = Maximum throughput per PDCH. If the network supports CS-1 to CS-4 channel coding, the maximum throughput will be the CS-4 rate.
>
> *Npdch* = Number of PDCHs in the cell.
>
> *Tflow* = Timer that manages the repetition of the flow control procedure.

R is more difficult to evaluate, as its value is dependent on the radio conditions on the air interface. It can be computed as follows:

$$R = MAXth * Npdch \qquad (6.3)$$

(the initial value when the BVC has just been established)

$$R = AVth * Npdch \qquad (6.4)$$

(the estimated value, taking into account the average throughput per time slot)

where

> *AVth* = Evaluated average throughput per time slot.

6.6.2 Downlink Transfer Procedure

This section describes how the transfer of downlink LLC frames within DL-UNITDATA PDUs on the Gb interface is managed when flow control at MS level is used in the BSS. An example of this procedure is described in Figure 6.33.

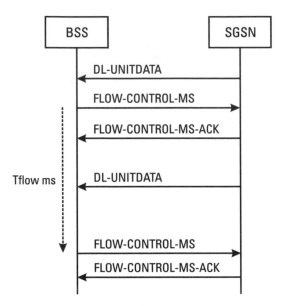

Figure 6.33 Downlink transfer procedure.

When the SGSN wants to send one DL-UNITDATA PDU that is addressed to one mobile for which no dedicated flow control parameters are available, it will use the default MS bucket parameters that are provided within the FLOW-CONTROL-BVC message.

When the BSS receives the DL-UNITDATA PDU, it starts the MS flow control procedure if it is supported by the BSS. The BSS creates a context for the new mobile and evaluates the bucket size and the bucket leak rate.

The bucket size B_{max} depends on the following parameters:

- The number of time slots (PDCH) that are allocated to the mobile for the downlink transfer;
- The maximum throughput that can be achieved per PDCH.

B_{max} can be computed as follows:

$$B_{max} = MAXth * Npdch\ ms * Tflow\ ms \tag{6.5}$$

where

Npdch ms = Number of PDCHs allocated to the mobile for downlink transfer;

MAXth = Maximum throughput per PDCH (if the network supports the channel coding scheme CS1 to CS-4, the maximum throughput will be CS-4 rate);

Tflow ms = Timer that manages the repetition of the flow control procedure.

The bucket leak rate *R* depends on the following parameters:

- The number of mobiles that are multiplexed in downlink on the same PDCHs as the ones allocated to the mobile;
- The average throughput on the PDCHs.

R can be computed as follows:

$$R = MAXth^* \sum_{1}^{Npdchms} 1/Nms \tag{6.6}$$

(the initial value when the BVC has just been established)

$$R = AVth * Npdch\ ms \tag{6.7}$$

(the estimated value, taking into account the average throughput per TS)

where

AVth = Evaluated average throughput per time slot;

Nms: = Number of mobiles multiplexed on the same PDCH.

6.6.3 Scenario for Cell Reselection During Downlink Transfer on Gb Interface

This section describes an example of a procedure that can happen when the mobile triggers a cell reselection during a downlink transfer. In this example, the cell reselection occurs between two cells belonging to the same SGSN.

During the downlink transfer, LLC frames that must be sent to the mobile are sent to the BSS in DL-UNITDATA PDUs. During the transfer on the radio interface of one LLC frame, a cell reselection is triggered, either by the network or by the mobile. In the case where it has been triggered by the network, a RADIO-STATUS PDU is sent to the SGSN indicating a "cell reselection ordered" cause. When the cell reselection is triggered by the MS, the network detects a radio interface problem and sends a RADIO-STATUS PDU after the reselection detection.

When the mobile arrives in the new cell, it triggers a cell update procedure. This procedure brings about the sending of an UL-UNITDATA PDU between the BSS and the SGSN that contains the cell update information. The SGSN deduces the new cell in which the mobile is located.

The SGSN then triggers a flush procedure in order to transfer the LLC frames that have not been sent on the air interface toward the new cell in the BSS. This can be the case if the two cells belong to the same NS entity and to the same RA. If not, or if the BSS is not able to transfer internally the LLC frames, they are removed from memory in the BSS. In this case, the SGSN will have to retransmit the deleted LLC frames toward the new cell in the BSS.

For an internal transfer of LLC frames within the BSS, the old cell and the new cell must belong to the same NS entity so that the BVC flow control context for the new cell in the SGSN can be updated. This is only possible by means of the flush procedure, which is internal to an NS entity. The two cells must also belong to the same RA so that the TLLI of the mobile is kept during the cell reselection. The BSS is able to start the downlink transfer of LLC frames on the air interface having the MS identifier.

Figure 6.34 describes the scenario during which the BSS transfers internally the LLC frames from the old cell to the new cell. After this transfer, the BSS sends a FLUSH-LL-ACK PDU in order to indicate that the LLC frames have been transferred and the amount of data impacted. This last parameter is used to adjust the BVC bucket parameter in the old cell context and the new cell context in the SGSN. Once the LLC frames are transferred toward the new cell in the BSS, their transfer on the air interface can start. As soon as the BSS is able to start the downlink transfer, it also triggers the MS flow control procedure in the new cell.

Figure 6.35 describes the scenario during which the BSS is not able to transfer internally the LLC frames from the old cell to the new cell. In this case, the BSS deletes the LLC frames within the old cell context and sends a FLUSH-LL-ACK PDU in order to indicate the amount of data that has

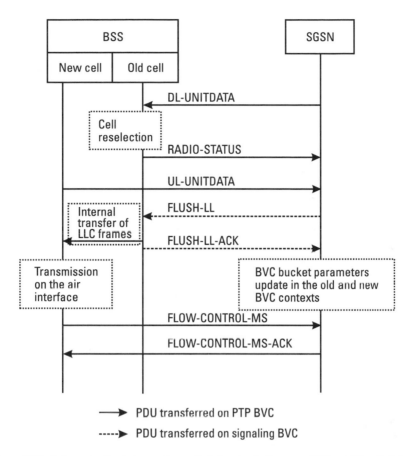

Figure 6.34 Cell reselection between two cells belonging to the same NSE and RA with internal transfer of LLC frames.

been deleted. This parameter is used by the SGSN to adapt the BVC bucket parameters in the old cell. The SGSN starts the retransmission of LLC frames that have not been sent on the air interface. At the reception of the first DL-UNITDATA PDU, the BSS starts the MS flow control procedure.

There is a great difference in the cell reselection interruption between these two procedures. In fact, in the first example (when the internal transfer of LLC frames is possible within the BSS), the downlink transfer interruption is shorter as the BSS is able to restart the retransmission of LLC frames as soon as the mobile has performed its cell update in the new cell. In the second scenario, the SGSN must retransmit the LLC frames to the BSS before being able to send on the air interface.

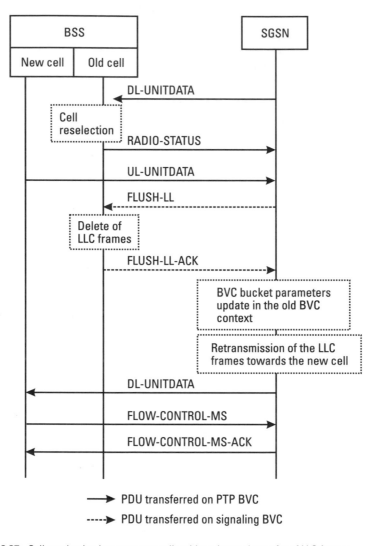

Figure 6.35 Cell reselection between two cells without internal transfer of LLC frames.

References

[1]   3GPP TS 08.16: Network Service (R99).

[2]   ITU-T recommendation Q.933 Annex A: Digital Subscriber Signaling System No. 1 (DSS 1)—Signaling Specifications for Frame Mode Switched and Permanent Virtual Connection Control and Status Monitoring.

[3]   Frame Relay Forum 1.1: User to Network Interface Implementation Agreement.

Selected Bibliography

3GPP TS 08.14: Gb Interface Layer 1 (R99).

3GPP TS 08.18: BSS GPRS Protocol (BSSGP) (R99).

3GPP TS 23.060: Service Description; Stage 2 (R99).

7

Signaling Plane

In a complex and distributed network architecture such as GPRS, some exchanges of information are required between the various nodes in order to control and support the transmission plane functions within the network. The necessary information used to coordinate the various nodes is called the signaling information. It is structured into messages.

This chapter covers the signaling related to mobility aspects and external packet network access. The focus in this chapter is on the signaling plane between the MS and the SGSN, and between the GSNs. The signaling related to RR management and the Gb interface are covered in Chapters 5 and 6, respectively.

One of the fundamental services in a GPRS network is mobility management. This must take into account functions such as access to GPRS services, paging for routing of incoming packet-switched calls, security, and location change of the GPRS subscriber. These functions are described in Section 7.1.

The signaling related to PDP context management enables the handling of access to an external packet-switching network via a GGSN. Routing information and QoS attributes characterize the PDP context and are negotiated during PDP context procedures. Section 7.2 describes PDP context management within the GPRS network.

The signaling between GSNs provides functions such as path management between two GSNs, tunnel management between an SGSN and a GGSN, location management between a GGSN and an GSN having an SS7 interface, and mobility management between two SGSNs. These functions

319

are provided by the GTP protocol in the control plane and are described in Section 7.3.

7.1   GMM

The management of GPRS mobility in the network ensures the continuity of packet services when a given subscriber moves from one GPRS LA to another. This implies that the network must know the identifier of the GPRS LA indicating where the MS is located.

The GMM functions enable the network infrastructure to keep track of subscribers' locations within the PLMN or within another PLMN. The SGSN, which is the serving node of an MS, handles the mobility context management related to it. This context contains information such as the IMSI, the P_TMSI, the RAI, and the CI. This mobility context management is also stored at the MS side, in the SIM card. All GPRS mobility procedures require a TBF connection at the RLC/MAC layer between the MS and the PCU.

7.1.1   Procedures

7.1.1.1   GPRS Attach Procedure

When an MS needs to signal its presence to the network in order to access to GPRS services, it performs an IMSI attach procedure for GPRS services. During this procedure a MM context is created between the MS and the SGSN.

There are two types of GPRS attach procedures:

1. *Normal GPRS attach.* This procedure is used by the MS to be IMSI attached for GPRS services only.

2. *Combined attach procedure.* This procedure is used by a class A or class B MS to be IMSI attached for GPRS and non-GPRS services in a cell that supports GPRS in network operation mode I (see Section 3.5.3.1).

Note that by default, the IMSI-attach procedure is referred to as the *attach procedure for circuit-switched services.* The IMSI-attach procedure for GPRS services is also called the *GPRS-attach procedure.*

Normal GPRS Attach

Figure 7.1 describes a GPRS-attach procedure. In this scenario, the MS signals itself to the network by sending it its old P-TMSI identifier associated with the old RAI identifier. When the SGSN receives this information, it analyzes the RAI identifier in order to determine the associated SGSN. If there is an SGSN change, the new SGSN must contact the old SGSN from its RAI identifier in order to retrieve the MS identity. Authentication functions may be performed; they are mandatory if no MM context information related to the MS, such as IMSI, P-TMSI, CI, and RA exists anywhere in the network. Then the new SGSN informs the HLR of SGSN change, and location information in the HLR database is updated via the MAP protocol on SS7 signaling. If the HLR receives an indication from an SGSN different from the one stored in its table for a GPRS subscriber, it requests the old SGSN to remove GPRS data related to this subscriber, and then transmits this data to the new SGSN.

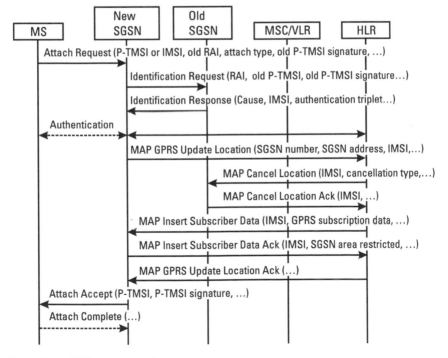

Figure 7.1 GPRS-attach procedure.

The new SGSN then transmits to the MS the GPRS-attach confirmation. If a new P-TMSI is allocated by the SGSN, it is acknowledged by the MS to end the GPRS-attach procedure.

Combined Attach Procedure

Figure 7.2 describes a combined GPRS/IMSI attach procedure in a cell that supports GPRS in network operation mode I. The difference with the previous scenario is that the new SGSN sends an IMSI attach request to the MSC/VLR via the Gs interface as soon as the new SGSN receives data related to the GPRS subscriber from the HLR. When the new SGSN receives the acceptance of IMSI attach from the MSC/VLR entity, it transmits the IMSI and GPRS attach confirmation. If the MS receives a new P-TMSI identifier or a new TMSI identifier, then it acknowledges it to end the combined attach procedure.

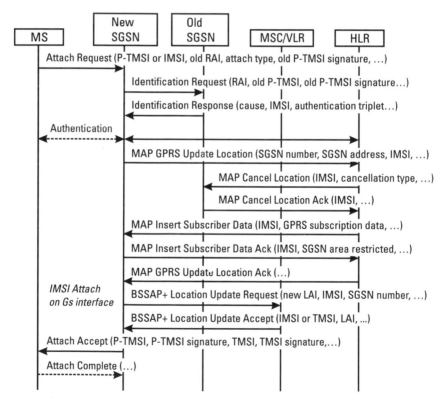

Figure 7.2 Combined GPRS/IMSI attach procedure.

7.1.1.2 GPRS Detach

When an MS does not need to access GPRS services anymore, an IMSI-detach procedure is initiated, either by the MS or by the SGSN. During this procedure, the MM context between the MS and the SGSN is removed.

There are two types of GPRS-detach procedures:

1. *Normal GPRS detach.* This procedure is used to IMSI detach only for GPRS services.

2. *Combined detach procedure.* This procedure is used to IMSI detach a class A or B MS for GPRS or non-GPRS services in a cell that supports GPRS in network operation mode I.

This procedure is initiated either by the MS or by the network

MS-Initiated Detach Procedure

MS-Initiated Normal GPRS Detach When a GPRS MS wishes to be IMSI detached for GPRS services, it initiates a GPRS-detach procedure to the SGSN. The procedure is ended upon the receipt of the DETACH ACCEPT message by the MS, as illustrated in Figure 7.3.

MS-Initiated Combined GPRS Detach When an MS both IMSI and GPRS attached wishes to perform a GPRS detach in a cell that supports GPRS in network operation mode I, it initiates a combined detach procedure to the SGSN. The latter sends an explicit request to the MSC/VLR to deactivate the association between SGSN and MSC/VLR in order that circuit-switched incoming calls are no longer routed to SGSN. Figure 7.4 illustrates this scenario. The same scenario is used for an MS both IMSI and GPRS attached wishing to be IMSI detached or both IMSI and GPRS detached in network operation mode I.

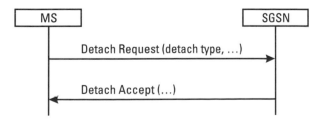

Figure 7.3 Normal GPRS detach initiated by MS.

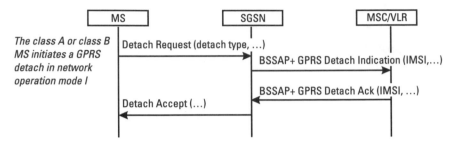

Figure 7.4 Combined GPRS detach initiated by an MS in network operation mode I.

Network-Initiated Detach Procedure

Network-Initiated Normal GPRS Detach When an SGSN wishes to IMSI detach a given MS for GPRS services, it initiates a GPRS-detach procedure. The procedure is ended upon the receipt of DETACH ACCEPT message by the SGSN, as illustrated in Figure 7.5. The network may request the MS to perform a reattach in the case of a network failure condition.

Network-Initiated Combined GPRS Detach When an SGSN wishes to IMSI detach a class A or B MS for GPRS or non-GPRS services, it notifies the relevant MS of a GPRS detach. It also sends an explicit request to MSC/VLR to deactivate the association between SGSN and MSC/VLR. Circuit-switched incoming calls are no longer routed to SGSN. Figure 7.6 illustrates this scenario.

An HLR may initiate a GPRS detach for operator purposes in order to remove the subscriber's MM and PDP contexts at the SGSN. The HLR sends a CANCEL LOCATION message in order to delete the subscriber's MM and PDP contexts from the SGSN. This latter then notifies the relevant

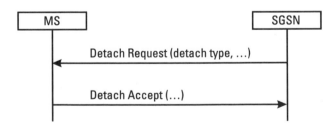

Figure 7.5 Normal GPRS detach initiated by an SGSN.

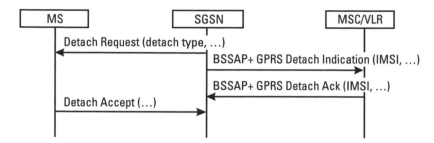

Figure 7.6 GPRS detach initiated by SGSN in network operation mode I.

MS of a GPRS detach. If the MS is both IMSI and GPRS attached, the SGSN sends an explicit request to the MSC/VLR to deactivate the association between the SGSN and the MSC/VLR. Figure 7.7 illustrates this scenario.

7.1.1.3 Paging on PCCCH

An MS, IMSI attached for GPRS services, may be paged on PCCCH channels if they are allocated in a cell it is camped. There are two types of paging on PCCCH:

1. GPRS paging in a cell that supports GPRS in network operating modes I or III;

2. Circuit-switched paging in a cell that supports GPRS in operating mode I (presence of Gs interface).

Note that paging modes are described in Section 3.5.3.1.

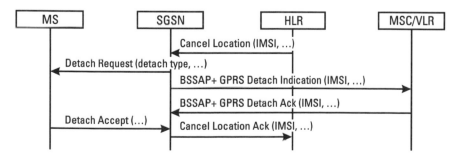

Figure 7.7 Combined GPRS detach initiated by HLR.

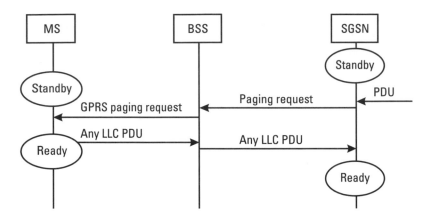

Figure 7.8   Packet-switched paging for network operation mode III with PCCCH channels.

GPRS Paging

In a cell supporting PCCCH channels and operating in network operation modes I or III, an MS both IMSI and GPRS attached in GMM STANDBY state can be paged by the network if the SGSN receives a PDU for it. When the MS receives a paging for packet-switched services, it requests of the BSS a TBF establishment in order to send any LLC frame to the SGSN. On LLC frame transmission, the MS goes to GMM READY state. Upon receipt of the LLC frame, the SGSN goes to GMM READY state. As the MS is now located at cell level, the packet transfer mode may start on downlink. This scenario is shown in Figure 7.8.

Circuit-Switched Paging

In a cell supporting PCCCH channels and operating in network operation mode I, an MS both IMSI and GPRS attached monitors the PCCCH channels in order to detect incoming circuit-switched calls. As soon as the MS recognizes its IMSI identifier in the GPRS PAGING REQUEST message received on the PPCH, it switches on the CCCH channels and requests the network to allocate radio resource in order to answer to circuit-switched paging. The MSC/VLR stops the paging procedure upon receipt of an SCCP connection establishment containing the paging response message from the MS via the A interface. This scenario is shown in Figure 7.9.

7.1.1.4   Authentication Procedure

The authentication procedure allows the network to identify and authenticate the user in order to protect the radio link from unauthorized GPRS calls.

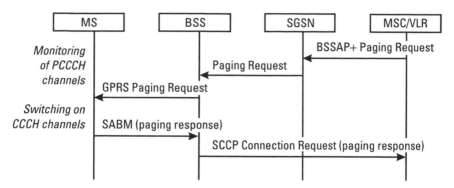

Figure 7.9 Paging for establishment of a circuit-switched connection for network operation mode I.

During an authentication procedure, a new SGSN needs to retrieve the triplet (Kc, signed RESult [SRES], random) from the HLR/AUC entity via the MAP protocol using the SS7 network. When the SGSN has retrieved this triplet, it authenticates the MS by sending the random number in the AUTHENTICATION AND CIPHERING request message. Upon receipt of this number, the MS will calculate the SRES number and the ciphering key Kc. Next the MS forwards to the network the SRES number. The network compares the SRES number calculated by the MS with the one sent by the entity HLR/AUC. If the two SRES numbers are identical, the SGSN considers that the outcome of the authentication of the GPRS subscriber is positive. Figure 7.10 illustrates the authentication of a GPRS subscriber.

Note that:

1. The GPRS ciphering algorithm is sent to the MS in the authentication and ciphering request message. The GPRS ciphering starts after the AUTHENTICATION AND CIPHERING response message is sent.

2. A GPRS *ciphering key sequence number* (CKSN) identifies the ciphering key Kc on the MS and network sides.

7.1.1.5 Location Procedures

A location procedure is always initiated by the MS. Under normal circumstances, a location change occurs when the MS decides to camp on a new cell for better radio conditions.

If an MS in GMM READY state camps in a new cell within its current RA, it needs to perform a cell update procedure in order to receive directly

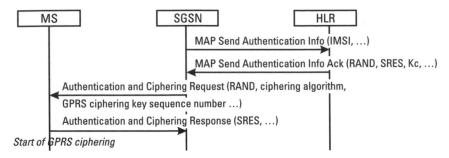

Figure 7.10 Authentication of a GPRS subscriber.

downlink PDUs from the network without being paged. If the MS camps in a new cell belonging to a new RA, it needs to perform an RA update procedure in order to update MM context information between the MS and the SGSN.

Cell Update

When a GPRS MS in GMM READY state detects a new cell within its current RA, it performs a cell update procedure by sending any LLC frame containing its identity. Figure 7.11 illustrates the cell update notification.

RA Update Procedure

An RA update procedure is performed when a GPRS MS has detected a new RA. This procedure is always initiated by the MS. There are four types of RA update procedures:

1. *Normal RA update*, performed by a class C MS or by a class A or B MS in a cell that supports GPRS in network operation mode II or III upon detection of a new RA;

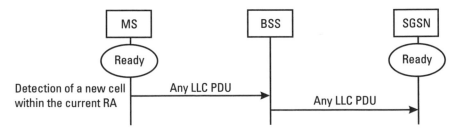

Figure 7.11 Cell update.

2. *Periodic RA*, performed by any GPRS MS upon expiry of a timer;

3. *Combined RA and LA update*, performed by a class A or B MS in a cell that supports GPRS in network operation mode I upon detection of a new LA;

4. *Combined RA and IMSI attach*, performed by a class A or B MS in a cell that supports GPRS in network operation mode 1 in order to be IMSI attached for non-GPRS services when the MS is already IMSI attached for GPRS services.

Normal RA Update

Intra-SGSN RA Update   During an RA update procedure, the MS signals itself to the SGSN by sending its old P-TMSI signature associated with the RAI identifier from its old RA. The SGSN has the necessary information about the MS if the SGSN also handles the old RA.

In the case of an intra-SGSN change, the SGSN validates the presence of the MS in the new RA by returning to it a ROUTING AREA UPDATE ACCEPT message. If the SGSN allocates a new P-TMSI identifier, it is acknowledged by the MS. This procedure is called intra-SGSN RA update since the SGSN does not need to contact an old SGSN, GGSN, and HLR. Figure 7.12 illustrates an intra-SGSN RA update procedure.

Note that a periodic RA update is always an intra-SGSN RA update procedure.

Inter-SGSN RA Update   When the SGSN detects that the old RA sent by the MS is handled by another SGSN, the SGSN has no information about this MS. In this case, the SGSN needs to contact the old SGSN, the GGSN, and

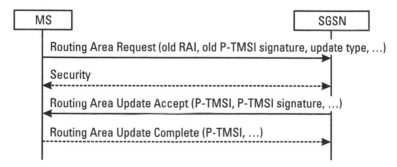

Figure 7.12   Intra-SGSN RA update.

the HLR in order to retrieve information and update the routing information. This procedure is called an inter-SGSN RA update procedure.

Thus the new SGSN is able to contact the old SGSN from the RAI identifier in order to retrieve MM and PDP context information related to the MS identified in the old SGSN by its old P-TMSI. If the old signature does not match the one saved in the old SGSN, the new SGSN performs an MS authentication procedure. If the old SGSN has saved in its buffer some packets addressed to the MS, it forwards the packets toward the new SGSN.

When the new SGSN has retrieved MM and PDP context information, it updates the data related to the new SGSN in the GGSN. The new SGSN then updates location information in the HLR database via the MAP protocol using the SS7 network. If the HLR receives an indication from an SGSN different from the one saved in its table for a GPRS subscriber, it requests the old SGSN to remove GPRS data related to this subscriber. It then transmits this data to the new SGSN.

When the new SGSN receives an RA update confirmation in the HLR database, it transmits the RA update confirmation to the MS with its new P-TMSI identifier and the receive N-PDU number. This message contains the acknowledgment of N-PDUs successfully transferred by the MS before the start of the update procedure. The RA update procedure ends as soon as the MS acknowledges its new P-TMSI identifier.

Figure 7.13 illustrates the inter-SGSN RA update procedure.

Combined RA and LA Update

During a combined RA and LA update procedure in a cell that supports GPRS in network operation mode I, the new SGSN (in case of SGSN change) retrieves MM and PDP context information related to the MS from the old SGSN. The new SGSN sends its address to the GGSN, updates routing information in the HLR via the MAP protocol, and retrieves data related to the GPRS subscriber from the HLR. The new SGSN transmits an LA update request via the Gs interface.

If the LA change involves a new MSC/VLR entity, the new MSC/VLR updates the location information in the HLR via the MAP protocol. When the HLR receives a notification from an MSC/VLR different from the one saved in its table, it requests the old MSC/VLR to remove data related to the GPRS subscriber and then transmits this data to the new MSC/VLR data.

When the new SGSN receives the LA update confirmation from the new MSC/VLR with the allocation of a new TMSI identifier value, it trans-

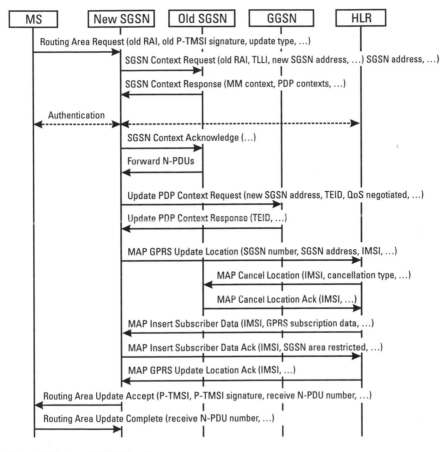

Figure 7.13 Inter-SGSN RA update.

mits the confirmation of the combined RA and LA update message toward the MS. The new SGSN allocates a new P-TMSI identifier for packet services and also returns the receive N-PDU number, containing the acknowledgment of N-PDUs successfully transferred by the MS before the start of the combined procedure. The combined RA and LA update procedure ends as soon as the MS acknowledges its new TMSI and P-TMSI identifiers.

Figure 7.14 illustrates the combined RA and LA update procedure.

Note that the combined RA and IMSI attach scenario generates the same message exchange between the MS, SGSN, GGSN, MSC/VLR, and HLR entities as the combined RA and LA update scenario.

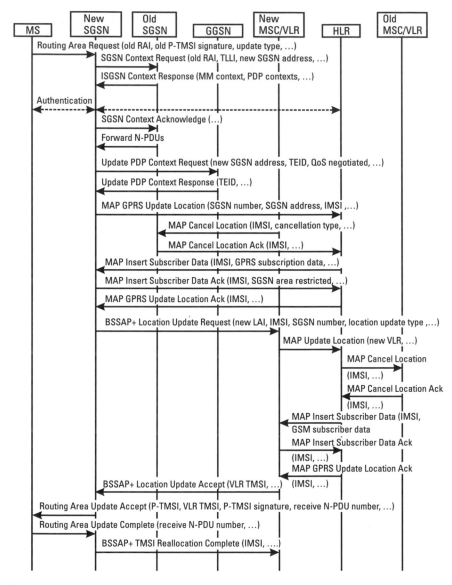

Figure 7.14 Combined RA and LA update procedure.

7.1.1.6 PLMN Selection

Selection Mode

When an MS that is both attached for GPRS and non-GPRS services exceeds the limits of PLMN radio coverage indicated by its subscription (see notion of home PLMN defined in Section 1.2.6), it may choose a different PLMN from its HPLMN (see notion of visited PLMN defined in Section 1.2.6).

There are two PLMN selection modes in the MS:

1. Automatic mode;
2. Manual mode.

In automatic mode, the MS attempts to register on a PLMN that belongs to a PLMN list sorted by priority order. In order to attempt a registration on a given PLMN, the MS camps on a cell for which the reception field power level meets an acceptable radio criterion. In case of registration failure on this PLMN, the MS registers on a PLMN that has the next best priority in its list. The MS scans its PLMN list from the highest down to the lowest priority until it succeeds in registering on a given PLMN.

In manual mode, the user selects manually the desired PLMN among the list of available PLMNs that is presented by the MS. The user has the opportunity to choose the PLMN selection mode at any time on the MS. At the power switch on, the default selection mode corresponds to the last used selection mode.

Management of Forbidden PLMNs

The MS receives an explicit registration reject on a given PLMN when the user subscription does not contain any roaming agreements with the operator of the *visited PLMN* (VPLMN). In this case the MS inscribes the PLMN identity in its forbidden PLMN list stored on its SIM card in order to avoid another registration request in automatic mode. This PLMN is removed from the list of forbidden PLMNs if the MS succeeds to register on it in manual mode.

As the roaming agreements between two operators may be different for circuit-switched services and packet-switched services, a given PLMN may reject a GPRS registration of a subscriber even when it may have already accepted a GSM registration of this subscriber. For this reason, the GPRS MS manages a specific list of forbidden PLMNs for GPRS in order to avoid another GPRS registration request. This list of forbidden PLMNs for GPRS is removed on SIM card extraction or at MS switch off. Moreover, a forbidden PLMN is removed from the list of GPRS forbidden PLMNs if the MS succeeds in registering on it in manual mode.

An MS configured in class A or B remains IMSI-attached for non-GPRS services in a network that operates in network operation modes II or III when it receives an explicit reject of GPRS subscription from the PLMN. When an MS configured in class A or B receives an explicit reject of GPRS subscription from the PLMN that operates in network operation mode I, it then performs another IMSI-attach procedure on the same PLMN. This is due to the fact that the IMSI-attach request has not been previously forwarded by the SGSN to the MSC/VLR. An MS configured in class C performs a new PLMN selection when it receives from the PLMN an explicit reject of GPRS subscription in automatic mode.

The reject cause "GPRS not allowed in PLMN" may also occur during an RA update procedure or during a combined RA and LA update procedure. In the case of an RA update procedure, an MS configured in class A or B in a network that operates in network operation mode II or III remains IMSI attached. In the case of a combined RA and LA update procedure a class A or B mobile performs a new LA update procedure in order to update its new LA. This is due to the fact that the location request has not been previously forwarded toward the MSC/VLR by the SGSN.

7.1.1.7   GPRS Parameter Update Status

The GPRS update status parameter is an internal MS parameter that gives the outcome of the last location request procedure such as GPRS attach or RA update request. This status defines the behavior of the MS. It is defined for a GPRS MS in any mode, when the GPRS function is disabled, when the MS is GPRS attached, or when the establishment of the GPRS-attach procedure is in progress. The service level to which a GPRS subscriber has access depends on the GPRS update status. This status is updated by a network-initiated GPRS attachment/detachment, authentication, and RA update procedure. The last value of this parameter is stored in the SIM card.

The values of these parameters are as follows:

- *GU1: UPDATED.* The GPRS update-status parameter is set to GU1 when the last attachment procedure or the last RA update procedure was successful. In this case, the SIM contains valid data related to the RAI, P-TMSI, GPRS ciphering key, and GPRS ciphering sequence number parameters.

- *GU2: NOT UPDATED.* The GPRS update-status parameter is set to GU2 when the last attachment procedure or last RA update procedure has failed (no response was received from the network after that the GMM procedure was initiatied by the mobile).

- *GU3: ROAMING NOT ALLOWED.* The GPRS update-status parameter is set to GU3 when the network response to a GPRS attachment or RA update procedure is negative. This case may occur on subscription or roaming restrictions. In this case, the SIM card does not contain valid data related to the RAI, P-TMSI, GPRS ciphering key, and GPRS ciphering sequence-number parameters.

7.1.1.8 Construction of the TLLI

The TLLI identifies a GPRS user. The relationship between TLLI and IMSI is known only in the MS and in the SGSN. The TLLI identifier is sent to the LLC layer in order to identify a logical link connection.

There are three types of TLLIs:

- *Local TLLI* is used when P-TMSI is valid—namely, when no location procedure is required;
- *Foreign TLLI* is used when an update of the P-TMSI is required by a location procedure;
- *Random TLLI* is used when no P-TMSI value exists.

Table 7.1
Structure of TLLI*

31	30	29	28	27	26 to 0	Type of TLLI
1	1	T	T	T	T	Local TLLI
1	0	T	T	T	T	Foreign TLLI
0	1	1	1	1	R	Random TLLI

*T bits are those derived from the P-TMSI value (length of 30 bits); R bits are chosen randomly by the MS when no P-TMSI is available.

The TLLI consists of 32 bits (its format is described in Table 7.1). The local TLLI value is deduced from the P-TMSI value allocated by the serving SGSN. Each time the P-TMSI value changes during a P-TMSI reallocation procedure, the GMM entity in the MS and the SGSN will rebuild the local TLLI value. The foreign TLLI value is deduced from the P-TMSI value allocated by an old SGSN. The random value is built directly.

7.1.2   Case Study: Control of the LLC Layer Operation

As the GMM service states reflect MS behavior, the GMM layer controls the following LLC operations:

- Sending or not sending user data;
- Sending or not sending signaling information.

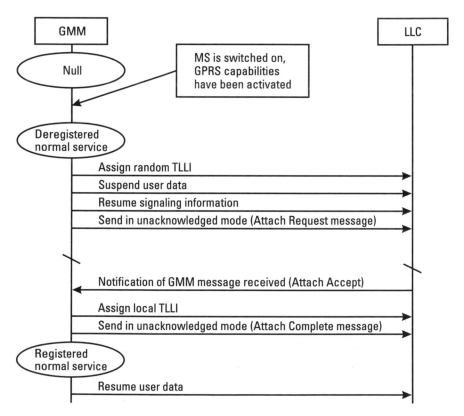

Figure 7.15 Control of LLC layer from GMM layer during a GPRS-attach procedure within the MS.

Figure 7.15 shows the exchange of primitives between GMM and LLC entities during a GPRS-attach procedure within the MS. When GPRS services are activated, the GMM entity will provide a random TLLI to the LLC entity if no valid P-TMSI is stored in the MS in order to initiate a GPRS-attach procedure. The signaling information flow is resumed at the LLC layer upon request from the GMM layer. As soon as a new P-TMSI is assigned by the SGSN, the GMM layer provides a new local TLLI to the LLC layer. At the end of the GPRS-attach procedure, the user data flow is resumed at the LLC layer upon request from the GMM layer.

Figure 7.16 shows the exchange of primitives between GMM and LLC entities during an RA update procedure within the MS. During this procedure, the user data flow is suspended in the LLC layer upon request from the GMM layer. A foreign TLLI is assigned to the LLC layer when the GMM layer detects a new RA in the selected cell. As soon as a new P-TMSI is

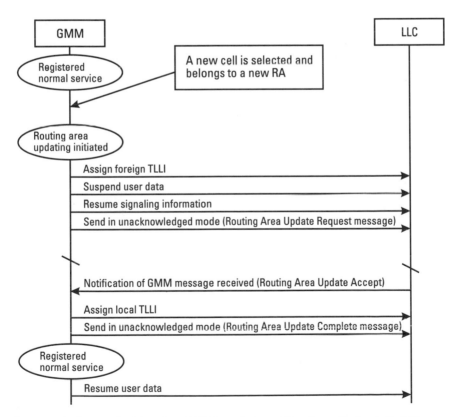

Figure 7.16 Control of LLC layer from GMM layer during an RA update procedure in the MS.

assigned by the SGSN, GMM provides a new local TLLI to LLC. At the end of the GPRS-attach procedure, the user data flow is resumed at LLC layer upon request from GMM.

7.2 PDP Context Management

A PDP context contains routing information for packet transfer between an MS and a GGSN to have access to an external packet-switching network. It is identified by an exclusive MS PDP address (mobile's IP address). This means that the MS will have as many PDP addresses as activated PDP contexts.

A concept of secondary PDP context has been introduced in order to have several PDP contexts sharing the same PDP address and the same access to the external packet-switching network. This concept was introduced for multimedia applications where each medium type requires specific transport characteristics and is mapped into a specific PDP context. It is based on the traffic flow template, which is a filtering mechanism used by the GGSN to route downlink IP packets toward the appropriate medium within the MS.

A given PDP context is in the active state when this PDP address is activated for data transfer. Before transferring data between an MS and a GGSN, it is necessary that a PDP context be activated.

PDP context procedures have been defined in order to create, modify, and delete PDP contexts within the MS, SGSN, and GGSN entities. The SM protocol is used between the MS and the SGSN and the GTP protocol is used in the controlling plane between the SGSN and the GGSN for PDP context procedures.

7.2.1 PDP Context Definition

A PDP context provides access to an external packet-switching network through the PLMN network. The data associated with the PDP context is as follows:

- *Access point name* (APN). This is the reference to a GGSN.
- *Network service access point identifier* (NSAPI). This is an index of the PDP context that uses the services provided by the SNDCP layer for GPRS data transfer. Up to 11 applications over the SNDCP layer may be identified by the NSAPI parameter. The NSAPI parameter is present in the SNDCP header.

- *LLC service access point identifier* (LLC SAPI). This identifies the SAP used for GPRS data transfer at the LLC layer.

- PDP address. This identifies the MS address related to a particular PDP context. This field consists of several fields including the PDP type (IP or PPP), PDP address type (IPv4 or IPv6), and address information containing the IP address.

- QoS. This defines the quality of service related to a particular PDP context. Parameters related to QoS are described in Section 2.4.

- Radio priority. This specifies the priority level used by the MS at the lower layers for transmission of data related to a PDP context.

- Protocol configuration options. This defines external network protocol options associated with a PDP context. It may contain information about protocols such as the *link control protocol* (LCP), the *PPP authentication protocol* (PAP), the *challenge handshake authentication protocol* (CHAP), and the *Internet Protocol Control Protocol* (IPCP).

Several PDP contexts can be activated at the same time in the MS. This means that the MS is able to transfer or receive data at the same time for several applications. Each PDP context is identified at the MS level, at the SGSN level, and at the GGSN level by the NSAPI. This identifier will be used to identify the logical link within the MS between the SNDCP entity and the application layer. An NSAPI is associated with an individual PDP address. When the SGSN receives a packet from the GGSN addressed to an MS identified by its PDP address, the SGSN inserts the associated NSAPI in the SNDCP header. Thus when the MS receives a packet from the SGSN, it identifies the appropriate application layer from the NSAPI parameter included in the SNDCP header.

7.2.2 PDP Address

A mobile's PDP address (IP address) can be assigned statically at the time of subscription or dynamically when the context is activated, depending on the operator's choice. A PDP address assigned permanently is called a static PDP address while a PDP address assigned during a PDP context activation is called a dynamic PDP address. The dynamic PDP address assignment is provided either by the GGSN, which creates a new entry in its PDP context table, or by the PDN operator.

7.2.3   Traffic Flow Template

In the GPRS Release 99 recommendations, several activated PDP contexts may share the same PDP address and the same APN. This is not the case in the earlier releases.

Some applications should require at the same time several activated PDP contexts with different QoS profiles while reusing the same PDP address and other PDP context information from an existing active PDP context. For example, multimedia applications involve several flows (e.g., voice and video that request different QoS but the same PDP address and the same APN). As a result of several PDP contexts activated at the same time with the same PDP address and the same APN, the different multimedia application flows can be routed between the MS and the GGSN via different GTP tunnels and possibly different LLC links.

This means that the analysis of PDP address destination cannot be used by GGSN to determine the NSAPI of the terminated application. In order to route downlink IP packets toward the terminated application within the MS, the GGSN uses a filtering mechanism called *traffic flow template* (TFT) that is defined by a set of packet filters. If several PDP contexts are associated with a PDP address, a TFT is created by the MS to specify an IP header filter for each or all but one context. This mechanism allows for the association of one packet filter with one NSAPI that is the identifier of the PDP context.

Each packet filter consists of a packet filter identifier within a TFT, a packet filter evaluation precedence that specifies the precedence for the packet filter among all packet filters in a TFT, and a list of packet filter attributes. Each packet filter attribute is deduced from IPv4 or IPv6 headers. The MS will define values related to each packet filter attribute. These may or may not be combined later in a packet filter. Each packet filter contains at least one of the following packet filter attributes:

- Source address and subnet mask—IPv4 or IPv6 address along with a subnet mask;
- Protocol number/next header—IPv4 protocol number or IPv6 next header value;
- Port numbers—port number or range of port number;
- Security parameter index—IPSec security parameter index;

- *Type of service* (TOS)/traffic class and mask—IPv4 TOS octet or IPv6 traffic class octet along with a mask;
- Flow label—an IPv6 flow label.

A TFT is created for a new PDP context using the same PDP address and the same APN as an existing PDP context but with a different QoS profile. This new PDP context is called a secondary PDP context and is activated during a secondary PDP context activation procedure. After a TFT has been created for a new secondary PDP context, it is sent by the MS to the network during the secondary PDP context activation procedure. A TFT may be modified during a PDP context modification procedure initiated by the MS. A TFT is deleted when the associated PDP context is deactivated.

During packet transmission between the MS and the external packet network, the GGSN will compare the parameters of the IP PDU header with packet filters of the TFT. If a match is found between the IP PDU header and a packet filter, the GGSN is able to direct the IP PDU from the interconnected external PDN to the suitable activated PDP context identified by the NSAPI parameter. This is illustrated in Figure 7.17.

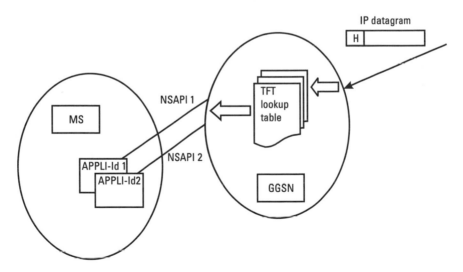

Figure 7.17 TFT filtering mechanism.

7.2.4 PDP State

A PDP context may or may not be activated for data transfer. This is indicated by the PDP state.

A PDP state set to INACTIVE means that the PDP context does not contain any routing information for packet transfer between an MS and GGSN for a given PDP address. Data transfer is not possible in this state. A PDP state set to ACTIVE means that a PDP context is activated in the MS, the SGSN, and the GGSN with a PDP address in use and routing information for packet transfer between the MS and the GGSN. Data transfer is possible in this state. The MS will be attached for GPRS services to be in ACTIVE PDP state.

Figure 7.18 shows the transition between the PDP states.

7.2.5 SM Layer

The SM layer handles the PDP context procedures such as PDP context activation, deactivation and modification, and secondary PDP context activation, between the MS and the SGSN.

The PDP context procedures handled by the SM layer can be performed for a given MS only if a GMM context has been established. Other-

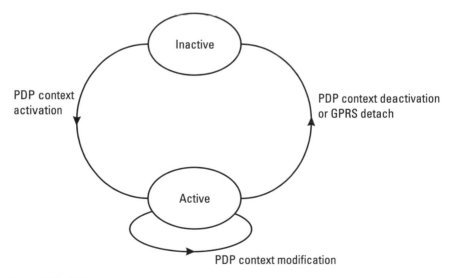

Figure 7.18 PDP states.

wise, the GMM layer must establish a GMM context for the use of the GPRS-attach procedure.

All PDP context procedures processed by SM peer entities require a TBF connection on the radio interface.

7.2.6   PDP Context Procedures

7.2.6.1   PDP Context Activation Procedure

The PDP context activation procedure may be initiated either by the MS or the GGSN.

Initiated by MS

When an MS wishes to create a PDP context, it sends a PDP CONTEXT ACTIVATION REQUEST message to the SGSN with optional parameters such as the requested QoS, requested NSAPI, MS PDP address, protocol configuration options, and APN. The requested NSAPI is provided by the MS among the ones not currently used by another PDP context in the MS. A PDP address is provided only if the MS already has a static address.

Security functions may be performed in order to authenticate the MS. The SGSN is able to derive the GGSN address from the APN identifier in order to forward this request to the GGSN. The SGSN creates a downlink GTP tunnel to route IP packets from the GGSN to the SGSN. The GGSN creates a new entry in its PDP context table to route IP packets between the SGSN and the external packet-switching network. The GGSN creates an uplink GTP tunnel to route IP-PDU from SGSN to GGSN.

The GGSN then sends back to the SGSN the result of the PDP context creation with the negotiated QoS and if necessary the MS PDP address. Next the SGSN sends an ACTIVATE PDP CONTEXT ACCEPT to the MS by returning negotiated QoS parameters, radio priority, and if necessary the MS PDP address.

Figure 7.19 illustrates a PDP context activation procedure initiated by the MS.

Initiated by GGSN

When the network receives an IP packet from an external network, the GGSN checks if a PDP context is already established with that PDP address. If not, the GGSN sends a PDU NOTIFICATION REQUEST to the SGSN in order to initiate a PDP context activation. The GGSN has retrieved the IP

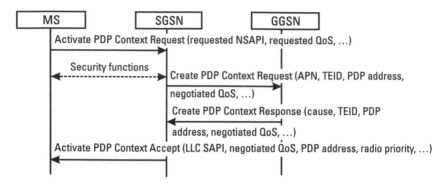

Figure 7.19 PDP context activation initiated by MS.

address of the appropriate SGSN address by interrogating the HLR from the IMSI identifier of the MS. The SGSN then sends to the MS a request to activate the indicated PDP context. Next the PDP context activation procedure follows the one initiated by the MS. Once the PDP context is activated, the IP packet can be sent from the GGSN to the MS.

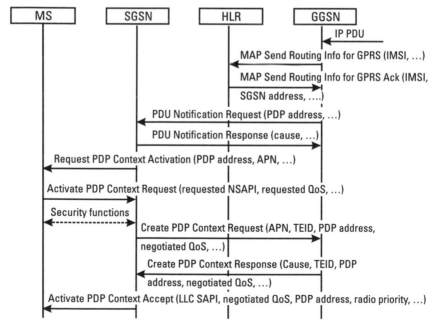

Figure 7.20 PDP context activation initiated by network.

Figure 7.20 illustrates a PDP context activation procedure initiated by the GGSN.

7.2.6.2 PDP Context Deactivation Procedure

As with the PDP context activation procedure, the PDP context deactivation procedure may be initiated by either the MS or the SGSN or GGSN.

Initiated by MS

In order to deactivate a PDP context, the MS sends a DEACTIVATE PDP CONTEXT message to the SGSN, which then sends a DELETE PDP CONTEXT message to the GGSN. If the PDP address was requested by the MS during the PDP context activation procedure, the GGSN releases this PDP address in order to keep it free for a subsequent PDP context activation. When the SGSN receives a PDP context deletion acknowledgment from the GGSN, the SGSN confirms to the MS the PDP context deactivation.

Figure 7.21 illustrates a PDP context deactivation procedure initiated by the MS.

Initiated by SGSN

When the PDP context deactivation is initiated by the SGSN, it sends a DELETE PDP CONTEXT REQUEST message to the GGSN. This procedure may occur when the HLR requests the deletion of PDP context due to the removal of general GPRS subscription data. The SGSN deletes the subscriber data when it receives the MAP DELETE SUBSCRIBER DATA message from the HLR. The SGSN acknowledges the delete subscriber data procedure by sending back a MAP DELETE SUBSCRIBER DATA ACK message to the HLR.

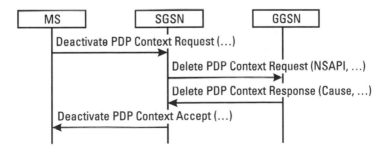

Figure 7.21   PDP context deactivation initiated by MS.

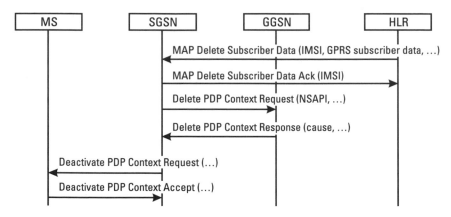

Figure 7.22 PDP context deactivation initiated by SGSN.

The GGSN deactivates the PDP context upon receipt of the DELETE PDP CONTEXT REQUEST message from the MS and releases the PDP address if this one was allocated dynamically during the PDP context activation procedure. When the SGSN receives a PDP context deletion acknowledgment from the GGSN, it initiates the PDP context deactivation in the MS by sending the DEACTIVATE PDP CONTEXT REQUEST message. When the MS has removed its PDP context, the MS sends back to the SGSN the DEACTIVATE PDP CONTEXT ACCEPT message.

Figure 7.22 illustrates a PDP context deactivation procedure initiated by the SGSN.

Iniated by GGSN

When the PDP context deactivation is initiated by the GGSN, it sends a DELETE PDP CONTEXT REQUEST message to the SGSN, which then sends to the MS a DEACTIVATE PDP CONTEXT REQUEST message. After having removed the PDP context, the MS sends a PDP context deactivation confirmation to the SGSN, which then sends a PDP context deletion acknowledgment to the GGSN. If the PDP address was requested by the MS during the PDP context activation procedure, the GGSN releases this PDP address.

Figure 7.23 illustrates a PDP context deactivation procedure initiated by the GGSN.

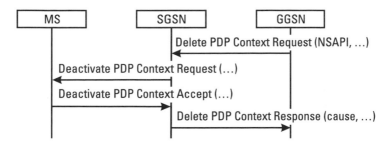

Figure 7.23 PDP context deactivation initiated by GGSN.

7.2.6.3 PDP Context Modification Procedure

The PDP context modification procedure is used to change the negotiated QoS, the radio priority level, or the TFT parameters negotiated during the PDP context activation procedure. This procedure may also be used to change the MS PDP address by the GGSN. The PDP context modification may be initiated by the MS, the SGSN, or the GGSN. The PDP context modification procedure initiated by the MS or GGSN was introduced in Release 99 of the GPRS recommendations, even though the PDP context modification procedure initiated by the SGSN was already introduced in Release 97 of the GPRS recommendations.

Initiated by SGSN

The PDP context modification initiated by the SGSN may occur after an inter-SGSN RA update procedure to change the negotiated QoS, the radio priority level, or the TFT negotiated during the PDP context activation procedure.

The SGSN sends an UPDATE PDP CONTEXT REQUEST message to the GGSN with a new QoS. The GGSN then checks if the new QoS is compliant with its capabilities and sends back to the SGSN in an UPDATE PDP CONTEXT RESPONSE message the negotiated QoS that takes into account if necessary some restrictions. The SGSN then sends new QoS parameters and a new radio priority parameter to the MS in a MODIFY PDP CONTEXT REQUEST message. If the MS accepts the new QoS parameters, it acknowledges the MODIFY PDP CONTEXT REQUEST message by sending to the SGSN a MODIFY PDP CONTEXT ACCEPT message.

Figure 7.24 illustrates a PDP context modification procedure initiated by the SGSN.

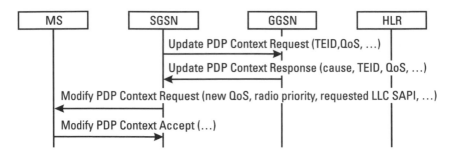

Figure 7.24 PDP context modification initiated by SGSN.

Note that if the new QoS is not accepted by the MS during the PDP context modification initiated by the SGSN, the MS will deactivate the PDP context by initiating the PDP context deactivation procedure.

Initiated by MS

The PDP context modification procedure initiated by the MS allows for a change in the negotiated QoS, the radio priority level, or the TFT negotiated during the PDP context activation procedure. The MS initiates the procedure by sending to the SGSN a MODIFY PDP CONTEXT REQUEST message, which may include a new requested QoS or new TFT parameters.

Next, the SGSN sends the new characteristics proposed for that PDP context to the GGSN by restricting if necessary the requested QoS in the UPDATE PDP CONTEXT REQUEST message. The GGSN then checks if the new QoS or TFT parameters are compliant with its capabilities and sends back to the SGSN in the UPDATE PDP CONTEXT RESPONSE message the negotiated QoS. When the SGSN receives the acknowledgment of the PDP context update from the GGSN, it sends to the MS a new radio priority and a packet flow ID on the QoS negotiated in a MODIFY PDP CONTEXT ACCEPT message.

Figure 7.25 illustrates a PDP context modification procedure initiated by the MS.

Initiated by GGSN

The GGSN initiates the procedure by sending to the SGSN an UPDATE PDP CONTEXT REQUEST message, which includes the desired QoS. A PDP address may be provided to the SGSN by the GGSN. Next, the SGSN sends to the MS the radio priority and packet flow ID on the requested QoS, which may be restricted if necessary in the MODIFY PDP CONTEXT

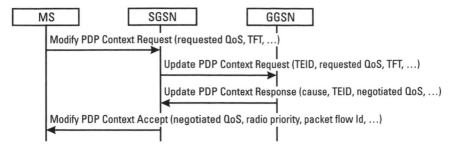

Figure 7.25 PDP context modification initiated by MS.

REQUEST message. The MS acknowledges the requested QoS by sending to the SGSN a MODIFY PDP CONTEXT ACCEPT message; the SGSN then sends this acknowledgment to the GGSN in the UPDATE PDP CONTEXT RESPONSE message.

Figure 7.26 illustrates a PDP context modification procedure initiated by the GGSN.

Note that if the new QoS is not accepted by the MS during the PDP context modification initiated by the GGSN, it initiates the PDP context deactivation procedure.

7.2.6.4   Secondary PDP Context Activation Procedure

When an MS wishes to create a secondary PDP context in order to reuse the same PDP address and the same APN, it sends a secondary PDP context activation request to the SGSN with the requested QoS and TFT. Security functions may be performed in order to authenticate the MS. The SGSN creates a downlink GTP tunnel to route IP packets from GGSN to SGSN. It then sends the requested QoS and TFT parameters to the GGSN in the CREATE PDP CONTEXT REQUEST message while indicating the

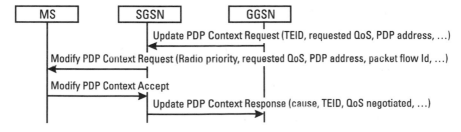

Figure 7.26 PDP context modification initiated by GGSN.

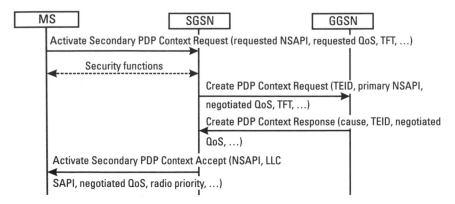

Figure 7.27 Secondary PDP context activation.

NSAPI assigned to the already activated PDP context with this PDP address.

The GGSN creates a new entry in its PDP context table to route IP packets between the SGSN and the external packet-switching network, which stores the TFT. The GGSN creates an uplink GTP tunnel to route IP-PDU from SGSN to GGSN. It then sends back to the SGSN the result of the secondary PDP context creation with a negotiated QoS. Next the SGSN sends an ACTIVATE SECONDARY PDP CONTEXT ACCEPT message to the MS by adding NSAPI and by returning QoS parameters and radio priority.

Figure 7.27 illustrates a secondary PDP context activation procedure initiated by the MS.

7.3 GTP Layer for the Control Plane

The GTP layer for the control plane (GTP-C) tunnels signaling messages between GSNs in the GPRS backbone network. The GTP-C enables several procedures to be performed through the GPRS backbone network, including path management, tunnel management, location management, and mobility management.

The path management is used by a GSN A to detect if a GSN B, with which the GSN A is in contact, is alive, or if a GSN has restarted after a failure. The tunnel management procedures are used to create, update, and delete GTP tunnels in order to route IP PDUs between an MS and an external PDN via the GSNs.

The location-management procedure is performed during the network-requested PDP context activation procedure if the GGSN does not have an SS7 MAP interface (i.e., Gc interface). It is used to transfer location messages between the GGSN and a GTP-MAP protocol-converting GSN in the GPRS backbone network.

The MM procedures are used by a new SGSN in order to retrieve the IMSI and the authentication information or MM and PDP context information in an old SGSN. They are performed during the GPRS attach and the inter-SGSN routing update procedures.

7.3.1 Path Management Procedure

The path management procedure checks if a given GSN is alive or has been restarted after a failure. In case of SGSN restart, all MM and PDP contexts are deleted in the SGSN, since the associated data is stored in a volatile memory. In the case of GGSN restart, all PDP contexts are deleted in the GGSN. A GSN A stores a GSN A restart counter in nonvolatile memory and a GSN B restart counter in volatile memory. After a GSN A restart, the GSN A restart counter is incremented, while the GSN-B restart counter is cleared. If a GGSN detects a restart in an SGSN, the GGSN deactivates all PDP contexts related to the SGSN. If an SGSN detects a restart in a GGSN, the GGSN deactivates all PDP contexts related to the GGSN before requesting the MS to reactivate them.

The path management procedure is activated by a GSN A toward a GSN B with which the GSN A is in contact. A GSN A sends an ECHO REQUEST message for each active UDP/IP path. This is considered as active when it is used to multiplex GTP tunnels between two GSNs for activated PDP contexts. If a GSN B receives an ECHO REQUEST message from a GSN A, it sends back an ECHO RESPONSE message with a recovery field that specifies its GSN restart counter. The GSN A compares its old GSN B restart counter with the one returned by the GSN B. If the two values do not match, the GSN A considers the GSN B to have been restarted. In this case, the GSN A considers that all PDP contexts used by the GSN B are inactive and therefore deletes these related PDP contexts.

Note that:

1. The event triggering the sending of an ECHO REQUEST message is implementation specific.

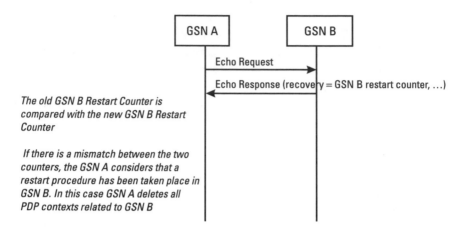

Figure 7.28 Path management procedure.

2. The GSN restart counter values are also exchanged between an
   SGSN and a GGSN during the PDP context procedures if the GSN
   A is in contact with the GSN B for the first time or if the GSN A
   has restarted and its GSN restart counter has not yet been sent to
   the GSN B.

Figure 7.28 illustrates the path management procedure.

7.3.2 Tunnel Management Procedures

Tunnel management procedures are defined to create, update, and delete
tunnels within the GPRS backbone network. A GTP tunnel is used to
deliver packets between an SGSN and a GGSN. A GTP tunnel is identified
in each GSN node by a TEID, an IP address, and a UDP port number.

A GTP tunnel is created during the PDP context creation subprocedure, modified during the PDP context update subprocedure, and deleted
during the PDP context removal subprocedure. All these subprocedures are
executed within the GSNs during the PDP context procedures.

During these procedures, a GSN provides two IP addresses to the peer
GSN: one for the control plane and another one for the user plane. These
GSN addresses are used by the peer GSN for the sending of messages in the
control plane and for the sending of packets in the user plane.

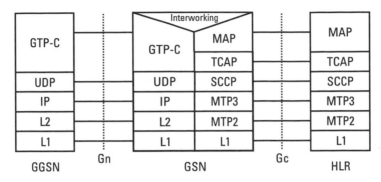

Figure 7.29 GGSN-HLR signaling via a GTP-MAP protocol-converter in a GSN. (*From:* [1].)

7.3.3 Location Management

Location management subprocedures are used between a GGSN that does not support an SS7 MAP interface (i.e., Gc interface) and a GTP-MAP protocol-conversing GSN. This GSN supports both Gn and Gc interfaces and is able to perform a protocol conversing between GTP and MAP. This is illustrated in Figure 7.29.

The location-management procedures are related to the network-requested PDP context activation procedure, and are used to determine both the SGSN IP address where the MS is located and whether the MS is reachable through this SGSN.

7.3.4 Mobility Management Between SGSNs

The MM procedures are used between SGSNs at the GPRS-attach and inter-SGSN routing update procedures.

An identity procedure has been defined to retrieve the IMSI and the authentication information in an old SGSN. This procedure may be performed at the GPRS attach.

A recovery procedure enables information related to MM and PDP contexts in an old SGSN to be retrieved. This procedure is started by a new SGSN during an inter-SGSN RA update procedure.

Reference

[1]   3GPP TS 29.060 GPRS Tunneling Protocol (GTP) Across the Gn and Gp Interface (R99).

Selected Bibliography

3GPP TS 03.60 Service Description; Stage 2 (R97).

3GPP TS 4.08 Mobile Radio Interface Layer 3; Stage 3 (R97).

3GPP TS 9.60 GPRS Tunneling Protocol (GTP) Across the Gn and Gp Interface (R97).

3GPP TS 23.060 Service Description; Stage 2 (R99).

3GPP TS 24.008 Mobile Radio Interface Layer 3 Specification; Core Network Protocols—Stage 3 (R99)

8

User Plane

The GPRS system must provide means of transmission in the user plane between the MS and external data packet network, according to the QoS related to a PDP context and the interface constraints of the PLMN. Despite the variety of interfaces across the GPRS network, the system must ensure an end-to-end transmission path.

This chapter deals with the user plane between the MS and the accessed external packet network. Protocols such as *subnetwork dependent convergence protocol* (SNDCP), LLC enable IP PDUs to be conveyed between the MS and the SGSN in acknowledged or unacknowledged modes. The user plane between the MS and SGSN is described in Section 8.1. As for the GTP layer line, it carries the IP PDUs within the GPRS backbone network. The user plane within the backbone network is described in Section 8.2. The user plane related to the air interface and Gb interface is covered in Chapters 5 and 6, respectively. Scenarios for IP sending within the GPRS PLMN are provided in Section 8.3.

A GPRS PLMN interconnects with an external data packet network. Interworking is described in Section 8.4.

8.1   Packet Transmission Between MS and SGSN

In the user plane between the MS and SGSN, the IP packets are processed by the SDNCP and LLC protocol layer entities. The SNDCP layer maps the IP protocol characteristics onto the characteristics of the underlying network. IP

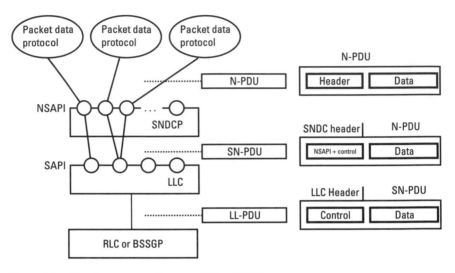

Figure 8.1   IP packets conveyed between MS and SGSN.

packets are conveyed in a *network protocol data unit* (N-PDU) and are encapsulated by the SNDCP layer into an *SNDCP PDU* (SN PDU) in order to be provided to the LLC layer. The LLC layer provides a reliable logical link between the MS and the SGSN. The LLC PDUs (LL PDUs) are provided by the LLC layer to the underlying layer and are encapsulated into RLC PDUs at air interface and into BSSGP PDUs at Gb interface.

Figure 8.1 illustrates the transport of IP packets between the MS and SGSN.

8.1.1   LLC Layer

8.1.1.1   General Description

The LLC layer provides logical links between a given MS and its SGSN. Several layer-3 protocols are located above LLC layer in the user plane such as SNDCP and GSMS, and in the control plane such as GMM and SM. The LLC layer operates above RLC layer at air interface and above BSSGP at Gb interface. The layer-three protocols use services provided by the LLC layer at the LLC layer SAP. Figure 8.2 shows the LLC layer structure.

A DLCI identifies a logical link connection. It consists of a *service access point identifier* (SAPI) and an MS's *temporary logical link identifier* (TLLI). The SAPI is used to identify the SAP between LLC layer and layer-three pro-

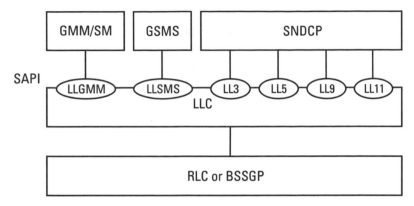

Figure 8.2 LLC-layer structure.

tocols above LLC. The TLLI is derived from the P-TMSI, and TLLI assignment is controlled by the GMM layer. TLLI is included in the LLC layer service primitives when an LLC frame is sent or received from lower layers (RLC/MAC layer or BSSGP layer).

The LLC layer supports various transmission modes, acknowledged and unacknowledged modes between the MS and the SGSN. In unacknowledged mode, the layer-three information is sent in numbered *unconfirmed information* (UI) frames, which are not acknowledged at the LLC layer. In acknowledged mode, the layer-three information is sent in order in numbered information (I) frames.

The GMM PDUs and SM PDUs use the unacknowledged transmission mode for LLC frames in the control plane as well as the SMS PDUs in the user plane. The SNDCP PDUs use either the unacknowledged or the acknowledged transmission mode for LLC frames in the user plane; the LLC transmission mode depends on QoS parameter values negotiated during the PDP context activation procedure.

SAPI1 (LLGMM) is used to identify the SAP of the LLC interface on the MS side and on the SGSN side with GMM/SM entity. SAPI7 (LLSMS) is used to identify the GSMS entity at the LLC-layer level. SAPI3, 5, 7, and 9 are used to identify the SNDCP logical entity at the LLC-layer level. Each SAPI for the SNDCP logical entity is configured differently depending on QoS parameters related to a given PDP context.

Note that a protocol discriminator field present in all GMM/SM messages allows for discrimination of GMM and SM messages sent on SAPI1 (LLGMM).

The LLC layer uses the concepts of link access procedure on the D-channel (LAPD) and is characterized by the following operations:

- *Error detection.* The presence of a FCS in LLC frame allows bit errors in the LLC frame header and information fields to be detected.

- *Error recovery and reordering.* This mechanism allows for the retransmission of unacknowledged frames in acknowledged transmission mode and the reordering of numbered acknowledged frames in order.

- *Flow control.* This is used in acknowledged transmission mode for congestion control of flows between the sender and the receiver when receiver does not have enough time to process the LLC PDUs received.

- *Window management for acknowledgment.* This is used to acknowledge a set of LLC frames in acknowledged transmission mode.

- Asynchronous balanced mode (ABM) operations. Such operations are used to establish or release a logical link in ABM mode in order to send LLC PDUs in an acknowledged transmission mode.

- *XID negotiation.* LLC- and SNDCP-layer parameters are negotiated at LLC-layer level between the sender and the receiver.

- *Multiplexing data.* A mechanism of SAPI-based LLC-layer contention resolution is used to multiplex several data flows from various upper layers.

Moreover, for data confidentiality, the LLC layer performs the ciphering function on LLC frames whenever the ciphering information (ciphering key Kc and ciphering algorithm) has been assigned to the LLC layer by the GMM layer. Only the information field and the FCS field are ciphered in LLC frames.

8.1.1.2 LLC Format Frame Description

The LLC frame format consists of a frame header (address field and control field), an information field, and an FCS field. The maximum number of octets in the information field (N201) is a LLC-layer parameter negotiated between the peer LLC entities during XID negotiation procedure. Table 8.1 gives the LLC frame format.

Table 8.1

LLC Frame Format

Bits

8	7	6	5	4	3	2	1

Address Field (1 octet)
Control Field
(variable length, max. 36 octets)
Information Field
(variable length, max. N201 octets)
Frame Check Sequence Field
(3 octets)

The SAPI is provided in the address field in order to identify the DLCI.

Table 8.2 lists the various LLC frames.

Note that an I frame is considered to be an I + S frame, since each I frame contains supervisory information.

The I + S information format frame contains the sequence number N(S) and the sequence number N(R). The N(S) sequence number indicates the send sequence number of transmitted I frames while the N(R) sequence number indicates that the LLC entity acknowledges all received I frames numbered up to and including N(R) − 1. The S information format frame contains only the transmitter receive sequence number N(R). The UI format frame contains the sequence number N(U), which is the unconfirmed sequence number of transmitted UI frames.

The FCS field contains a 24-bit *cyclic redundancy check* (CRC) code. For the I frames, the CRC calculation is performed over the entire contents of the frame header and the information field. For the UI frames, the LLC layer offers two protection modes, unprotected mode and protected mode.

Table 8.2
LLC Frames

LLC Frame Format	Command	Response	Meaning
Information format frame	I	I	Information
Unacknowledged format frame	UI		Unacknowledged Information
S Supervisory format frames	RR	RR	Receiver Ready
	ACK	ACK	Acknowledgment
	SACK	SACK	Selective Acknowledgment
	RNR	RNR	Receiver Not Ready
U Unnumbered format frames	SABM		Set ABM
		UA	Unnumbered Acknowledgment
	DISC		Disconnect
		DM	Disconnected mode
		FRMR	Frame Reject
	XID	XID	Exchange Identification
	NULL		Null

In protected mode, the CRC is calculated on the entire frame. In unprotected mode, the CRC is calculated on the LLC frame header and on the SNDCP PDU header.

Note that the protection mode is defined during the PDP context activation procedure.

8.1.1.3 Acknowledgment Window Mechanism

The window size for acknowledgment is configured during the XID negotiation procedure and may be up to 255. It means that a sender may send a number of successive LLC frames equal to the window size in acknowledged transmission mode without waiting for the receiver's acknowledgment of the LLC frames already sent.

Note that a small window size implies that the receiver sends acknowledgments often and the sender waits for them often to transmit new LLC frames. A large window size implies a large buffer in the sender side in order to contain all LLC frames sent but not acknowledged by the receiver. The

the window size is therefore configured according to the availability of memory in the sender.

In order to manage the windows for transmission and reception, the sender and the receiver use several internal variable states. The sender manages a variable state V(S), which indicates the next in-sequence I frame to be transmitted and a variable state V(A), which indicates the last LLC frame sent and not yet acknowledged by the receiver. The sender numbers each I frame sent to the receiver with a send sequence number N(S), which is set to V(S). The receiver manages a variable state V(R), which indicates the sequence number of the next in-sequence I frame expected to be received. V(R) is incremented by one upon receipt of in-sequence I frame, whose send sequence number N(S) is equal to V(R). All I frames and supervisory frames from the receiver contain the expected send sequence number of the next in-sequence received I frame N(R), which is set equal to V(R) to indicate to the sender that all I frames numbered up to and including N(R) − 1 have been correctly received. The sender updates V(A) by N(R) received from the receiver.

Whenever an LLC entity receives an I + S frame, it analyzes an acknowledgement request bit (A bit) in the control field. If acknowledgement is requested by the sender (A bit set to 1), then the receiver will acknowledge positively or negatively each LLC frame received in an S or I + S frame. The sender will retransmit all I frames not acknowledged positively by the receiver.

The receiver sends an acknowledgment with S or I + S frames under the following form:

- Acknowledgment of previously received I frames numbered up to and including N(R) − 1 (I frame RR format or I frame RNR format or S frame RR format or S frame RNR format);

- Acknowledgment of previously received I frames numbered up to and including N(R) − 1, and frame N(R) + 1 (I + S frame ACK format or S frame ACK);

- Acknowledgment of all previously received I frames numbered up to and including N(R) − 1 and frames indicated by the SACK bitmap (I frame SACK format or S frame SACK).

Note that if the sender receives an I frame ACK format or S frame ACK format, the sender will retransmit only the frame number N(R). If the

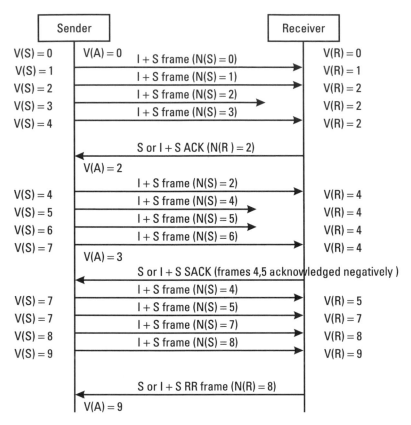

Figure 8.3   Example of state variable management .

sender receives an I frame SACK format or S frame SACK format, the sender will retransmit all I frames not acknowledged in the SACK bitmap.

Figure 8.3 gives an example of state variable management.

8.1.1.4   Data Multiplexing

The LLC layer can manage several logical link connections simultaneously. A *logical link entity* (LLE) controls one logical link connection. As a logical link connection is identified by a DLCI consisting of SAPI and TLLI as there are different SAPIs, each connection is managed by an LLE.

On frame transmission, each LLE prepares the LLC format on each frame received from the upper layers by completely filling out the address field and the information field of the LLC frame, and partially filling out the control field with the appropriate command or response. The multiplex procedure allows several flows from several LLEs and priorities between the various LLEs to be managed using a SAPI-based LLC-layer contention resolution. The multiplex procedure ends after the control field of the LLC frame is filled. Then it inserts an FCS, performs the ciphering function, and sends the LLC frame to RLC/MAC layer on MS side or to BSSGP layer on SGSN side.

On frame reception, the multiplex procedure, which receives a frame from RLC/MAC on MS side or from BSSGP on SGSN side along with the TLLI parameter, performs the deciphering function, verifies the CRC in the FCS field, and finally analyzes the SAPI in the address field in order to route the LLC frame to the appropriate LLE (identified by DLCI). The LLE extracts the information field from the LLC frame in order to send it to the upper layer across the appropriate SAPI. In acknowledged mode, the LLC guarantees in-order delivery to the upper layers.

Figure 8.4 illustrates the multiplexing procedure.

8.1.1.5 XID Negotiation

This procedure consists in the exchange of some SNDCP- and LLC-layer parameters on SAPIs assigned to a layer-three entity. The negotiated parameters may be at SNDCP level parameters such as header compression and data compression information, and at LLC level parameters such as maximum number of retransmission in acknowledged transmission mode, maximum information field length for I frames or U or UI frames, ciphering input offset value for UI or I frames and retransmission timer value. This procedure may occur at the end of a PDP context activation procedure or at an SGSN change during the establishment of ABM. The XID negotiation procedure is performed by the LLC upon request from the SNDCP layer. The XID negotiation is a one-step procedure. The initiator proposes a set of values for LLC and SNDCP parameters within the allowed values in an I + S frame XID format or an S frame XID format. The receiver may confirm these values or offer other values for these parameters.

8.1.2   SNDCP Layer

The purpose of the SNDCP layer is to multiplex data coming from different PDP sources (e.g., IP, PPP) in order to be sent across the LLC layer in acknowledged or unacknowledged transmission mode. All N-PDUs from the network layer protocols above the SNDCP layer will be carried out in a transparent way through the GPRS network.

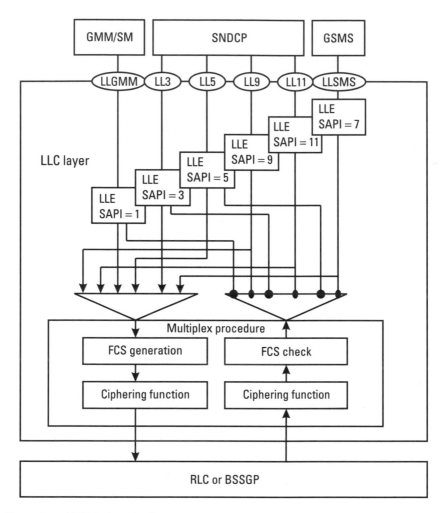

Figure 8.4 Multiplexing procedure.

In order to improve channel efficiency, the SNDCP entity provides optional compression features on N-PDUs for protocol control information (TCP/IP header compression) and for data. Segmentation and reassembly procedures are performed by the SNDCP entity when the SNDCP PDU exceeds a given length (N201).

Table 8.3
SN-DATA PDU Format

Bits

Octets	8	7	6	5	4	3	2	1
1	X	F	T	M	NSAPI			
2	DCOMP				PCOMP			
3	N-PDU number–acknowledged mode							
N	Data segment							

8.1.2.1 SNDCP Format Frame Description

Two different SN-PDUs are defined, the SN-DATA PDU for acknowledged data transfer and SN-UNITDATA PDU for unacknowledged data transfer. Each SN-PDU format consists of a header part and data part.

The SN-DATA PDU format is given in Table 8.3.

The SN-UNITDATA PDU format is given in Table 8.4.

Table 8.4
SN-UNITDATA PDU Format

Bits

Octets	8	7	6	5	4	3	2	1
1	X	F	T	M	NSAPI			
2	DCOMP				PCOMP			
	Segment number				N-PDU number—unacknowledged mode			
3	N-PDU number—acknowledged mode							
4	N-PDU number—unacknowledged mode (continued)							
N	Data segment							

The fields of the SNDCP PDU header have the following meaning:

- M: flag that indicates if this is the last N-PDU segment (used for segmentation and reassembly of N-PDUs);
- T: flag that indicates if this is a SN-DATA PDU or a SN-UNIT-DATA PDU;
- F: flag that indicates if this is the first segment of an N-PDU (used for segmentation and reassembly of N-PDUs);
- X: spare bit;
- DCOMP: data compression identifier if there is data compression (only present for the first segment);
- PCOMP: protocol control information identifier if there is header compression (only present for the first segment);
- Segment number: sequence number of SN-UNITDATA PDU (due to the unreliable nature of unacknowledged transmission mode);
- N-PDU number—acknowledged mode: N-PDU number of an N-PDU sent in acknowledged transmission mode (only present for the first segment);
- N-PDU number—unacknowledged mode: N-PDU number of an N-PDU sent in unacknowledged transmission mode.

8.1.2.2 Multiplexing of N-PDUs

One PDP may have several PDP contexts identified by an NSAPI. The NSAPI allows the upper source above SNDCP entity to be identified; it is contained in each header of SNDCP PDU (SN-PDU).

When the SNDCP entity receives an N-PDU from the upper layers on a given NSAPI, it encapsulates the N-PDU into a SN-PDU by updating the NSAPI value in the header, and then sends it to the appropriate SAPI of the LLC layer. When the SNDCP entity receives an SN-PDU from the LLC layer on a given SAPI, it deencapsulates the N-PDU in the SN-PDU and routes it to the upper layer according to the NSAPI contained in the SN-PDU.

The NSAPI (11 possible values) and LLC SAPI (4 possible values) are assigned to the SNDCP entity by both the SM entity in the MS and in the SGSN for a given PDP context at the end of the PDP context activation procedure or PDP context modification procedure. An LLC SAPI may be shared by several NSAPIs.

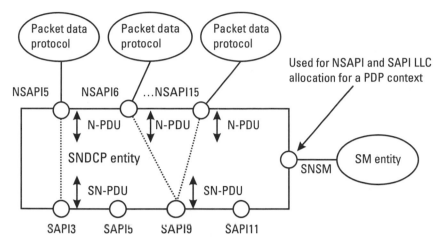

Figure 8.5 Multiplexing of N-PDUs.

Figure 8.5 shows the multiplexing of N-PDUs.

8.1.2.3 Establishment and Release of Acknowledged LLC Operation

The SNDCP entity initiates the ABM establishment procedure on a given SAPI if an LLC acknowledged transmission mode is required. At the end of this procedure, the given SAPI is in ABM mode.

The SNDCP entity initiates the ABM release procedure on a given SAPI if there is no NSAPIs requiring an LLC acknowledged transmission mode on this SAPI. It may occur during a PDP context deactivation procedure if the NSAPI was mapped to this SAPI or during a PDP context modification procedure if there is a different mapping between the NSAPI and the SAPI. At the end of this procedure, the given SAPI is in *asynchronous disconnected mode* (ADM).

Note that the LLC transmission mode is defined during the PDP context activation procedure.

Figure 8.6 shows several scenarios for ABM establishment and release procedures. The first part of the scenario shows an establishment of ABM mode on SAPI3 upon receipt of an indication from the SM entity specifying a PDP context activation with the use of SAPI3 by the allocated NSAPI1. The second part of the scenario shows a release of ABM mode on SAPI3 and an establishment of ABM mode on SAPI5 upon receipt of an indication of a PDP context modification from the SM entity requiring that the previous NSAPI be mapped to SAPI5. The third part of the scenario shows the ABM

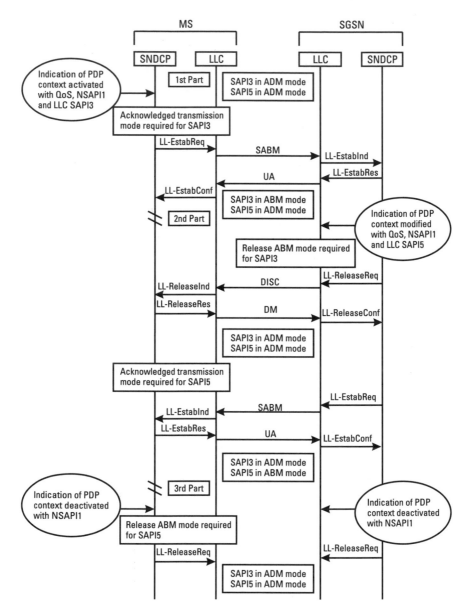

Figure 8.6   Scenarios for ABM establishment and release procedures.

release procedure on SAPI5 upon receipt of indication of a PDP context deactivation from the SM entity.

8.1.2.4 Protocol Control Information Compression

The SNDCP entity may propose an optional feature, a protocol control information compression. This is applied on the N-PDU header. There are two compression algorithms supported by the SNDCP layer for protocol control information, the RFC-1144 algorithm (TCP/IP header compression) and the RFC-2507 algorithm (TCP/IP and UDP/IP header compression). A protocol-control-information-compression entity may be shared by several NSAPIs.

The parameters associated with a protocol control information compression are negotiated between protocol-control-information-compression entities within the SNDCP layer in the MS and in SGSN during the XID negotiation procedure.

8.1.2.5 User Data Compression

The SNDCP entity proposes an optional feature for user data compression. The latter may apply on the entire N-PDU for acknowledged and unacknowledged transmission modes and is performed after protocol control information compression if it is used. The only algorithm used for data compression is V.42 bis.

The V.42 bis data compression uses two dictionaries for one direction transmission, one dictionary managed by the data-compression encoder and the other by the data-compression decoder. A dictionary stores strings for use in the encoding and decoding process represented by a set of trees, a hierarchical data structure. A tree consists of nodes; each node corresponds to a character in the alphabet. A node with no parent node is the hierarchically higher level in the tree, called the root node, and represents the first character of a string. A node with no dependent node is the hierarchically lower level in the tree, called the leaf node, and represents the last character of a string. Each node is identified by a unique codeword within the encoder dictionary and the decoder dictionary.

Figure 8.7 shows the tree structure.

The encoding function performs a string-matching procedure by comparing a sequence of characters (data flow received) with a dictionary entry. In case of matching, the string is sent in compressed mode and is encoded with the codeword associated with it. In case of no matching, the string is sent in transparent mode; the encoder adds the new string in its dictionary

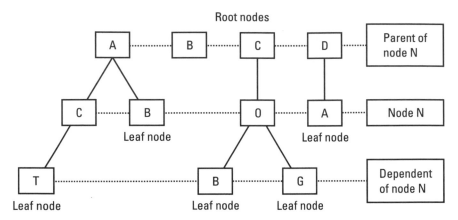

Figure 8.7   Tree structure.

with a new node onto a tree by appending a new character to an existing string. A codeword is associated with the next empty dictionary entry; its value is incremented for each new entry. Each time there is a transition of mode of operation from compressed mode to transparent mode or vice versa, the encoder function indicates to the decoder function the transition of encoder state.

The decoding function operates in both compressed and transparent modes. In compressed mode, the decoder retrieves the original sequences of characters identified by codewords in its dictionary. In transparent mode, the decoder adds the new string received in its dictionary by adding a new node onto the tree. A codeword is associated with the string in a manner consistent with the encoding functions.

A V.42 bis compression entity may be shared by several NSAPIs if they use the same dictionary, but there are two separate V.42 bis compression entities for SN PDU in acknowledged and unacknowledged transmission modes. The parameters associated with the V.42 bis data-compression function are negotiated between the data-compression entities within the SNDCP layer in the MS and in SGSN at link establishment via the XID negotiation procedure. These parameters allow for the configuring of the direction of the compression (no compression, or compression from MS to SGSN, or compression from SGSN to MS, or compression for both directions), the maximum number of codewords in the dictionary and the maximum characters in a transparent mode, and the applicable NSAPIs.

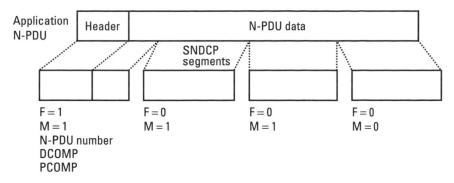

Figure 8.8 Segmentation mechanism for acknowledged transmission mode.

8.1.2.6   Segmentation and Reassembly

The SNDCP entity segments an N-PDU into several SN-PDUs to avoid having the SN-PDU transmitted across the SAPI LLC exceed a given length N201. This latter is negotiated by LLC entities during the XID negotiation procedure. There are two N201 values, N201-I value for acknowledged transmission mode and N201-U for unacknowledged transmission mode.

The segmentation reassembly operation uses the flags M and F in order to delimit an N-PDU. The DCOMP and PCOMP will be included in the header only for the first SN-PDU (F bit set to 1). For acknowledged transmission mode, the N-PDU number is also present in the header of the first SN-PDU. For unacknowledged transmission mode, the segment number field is incremented by one for each SN-UNITDATA PDU.

Figure 8.8 shows the segmentation mechanism for acknowledged transmission mode. Figure 8.9 shows the segmentation mechanism for unacknowledged transmission mode.

8.1.3   Case Study: Buffer Management

A buffer management will be put in place for data processing on uplink and on downlink. In order to define a strategy for global buffer management, all functional constraints related to data processing will be identified to determine if the data processing is performed in a new buffer or in the same buffer used by the previous data processing. In one case there is a buffer recopy between each new data processing, while in the other case a simple pointer on the shared buffer is passed. The last solution enables better performance,

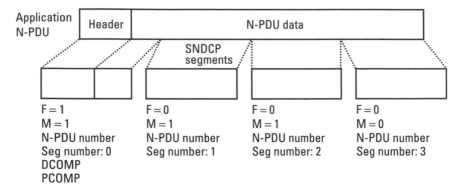

Figure 8.9   Segmentation mechanism for unacknowledged transmission mode.

since there is no time wasted for data recopy from one buffer to another one. In addition, there is an obvious interest in minimizing the number of used buffers for data processing in order to optimize the memory usage, especially for MSs. But as the data transfer processing is performed across several protocol layers from IP layer to RLC/MAC and vice versa, the complexity of data processing algorithms increases if they are all processed in the same buffer. Moreover, the level of interface opening between the protocol layers decreases, since each protocol layer is more dependent on the other.

The buffer management strategy takes into account a priority management when several traffics are processed in the same time. The memory management may define either a memory resource shared by all traffic or a separated memory resource for each traffic.

There is a tradeoff between the various aspects such as performance, optimization of memory use, data processing complexity, and level of interface opening during software design for data transfer processing. This tradeoff is difficult to find because of the transmission chain complexity. In some cases, a simulation system tool may be useful to evaluate the transmission chain complexity, since it enables estimation of the impacts on the transmission chain performance of several configuration parameters, such as buffer length, number of buffers, length of file, number of files, and multiplexing of several traffics in parallel.

8.1.3.1   Examples of Buffer Usage from SNDCP Layer to LLC Layer

A strategy for buffer management from the SNDCP layer to the LLC layer takes into account the various examples of data processing seen in the previous section: protocol control information compression, user data compression, N-PDU segmentation into several SN-PDUs, data multiplexing, addition of LLC headers, addition of CRC, and ciphering of the LLC frame. More functional constraints, such as the following, are also taken into account in this strategy:

- Waiting for acknowledgment (at LLC or SNDCP layer level);
- Internal flow control between protocol layers;
- Segmentation of N-PDU into several SN-PDUs according to N201 length value negotiated by CID negotiation;
- QoS requirements on each SAPI LLC and on each NSAPI SNDCP.

Two examples of buffer management from the SNDCP layer to the LLC layer are described below, one utilizing a memory recopy and the other without memory recopy. There are other solutions that combine the two solutions.

Example 1: Recopy from SNDCP Layer to LLC Layer

The SNDCP copies the result of IP header compression and of the user data compression in two new buffers. When SNDCP segments one N-PDU into several SN-PDUs, it allocates a new buffer for SN-PDU with its header part and data part by reserving a memory length equal to the length of the LLC frame in order that the LLC layer should append to the received SN-PDU an LLC header and a CRC in the SN-PDU buffer. The LLC copies the result of the ciphering process on the LLC frame into a new buffer.

Figure 8.10 illustrates the buffer recopy mechanism from the SNDCP layer to the LLC layer.

The implementation of this solution is very simple. It provides for independence between each layer, since data processing is performed by each layer in its own buffer. This solution has an impact on performance in terms of time and on memory consumption due to memory recopy between each data processing layer.

Example 2: No Recopy from SNDCP Layer to LLC Layer

The SNDCP updates in the memory the previous received IP frame with the compressed IP header. No user data compression is performed in this example. When the SNDCP segments one N-PDU into several SN-PDUs, it allocates a new buffer containing the list of header parts associated with each

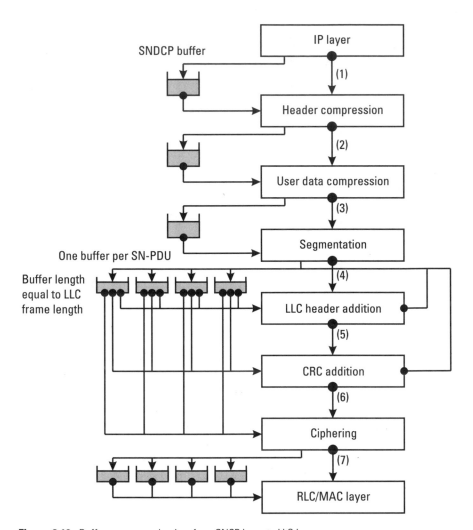

Figure 8.10 Buffer recopy mechanism from SNCP layer to LLC layer.

SN-PDU, and each SN-PDU data part is addressed by a pointer in the
N-PDU buffer. Next, the SNDCP layer provides a list of two chained pointers to the LLC layer, the first one pointing on the N-PDU buffer at a given
place for the SN-PDU data part, the second one pointing on a given place of
the SN-PDU header buffer. In order to add an LLC header and CRC to an
SN-PDU, the LLC allocates two new buffers, one buffer containing a list of
LLC headers associated with each LLC PDU and another one containing a

list of CRCs associated with each LLC PDU. Then a list of four chained pointers is provided at the input of the ciphering function, the first one pointing on N-PDU buffer, the second one pointing on SN-PDU header buffer, the third one pointing on LLC header buffer, the fourth one pointing on CRC buffer. Next, memory contents located in the four buffers that are identified by the previous list of four chained pointers are replaced with the result of the ciphering operation.

Figure 8.11 illustrates the second example with no buffer recopy from the SNDCP layer to the LLC layer.

This solution implies better performances in terms of time due to less memory recopy between each data processing event and less memory consumption, for the same reason as that stated above. With this solution, the layer entities are more dependent from the others. For example, the SNDCP layer must wait for all processing performed by the lower layers such as LLC and RLC to end before preparing a new N-PDU. The implementation of this solution is complex, since all data processing is based on a list of chained pointers. It may be difficult to tune and may be a source of potential bugs. Moreover, it may pose problems with respect to porting this solution from one system to another.

Table 8.5 lists the advantages and the drawbacks with respect to the implementation of the two solutions for the memory usage from the SNDCP layer to the LLC layer.

8.1.3.2   Examples of Buffer Usage from LLC Layer to SNDCP Layer

A strategy of buffer management from the LLC layer to the SNDCP layer takes into account the various data processing events such as deciphering of the LLC frame, CRC checking, data demultiplexing, reassembly of several SN-PDUs into N-PDU, user data decompression, and protocol control information decompression. An acknowledgment mechanism and peer-to-peer flow control is taken into account at the LLC-layer level for data receipt in acknowledged mode. In this case, the LLC layer ensures the reordering of the received LLC frames and makes easier the reassemby procedure. While in unacknowledged mode, the SNDCP layer ensures the reordering and the full reassembly procedure, since it is not performed by the LLC layer.

Two examples of buffer management from the LLC layer to the SNDCP layer are described below, one example with a memory recopy and the other without memory recopy.

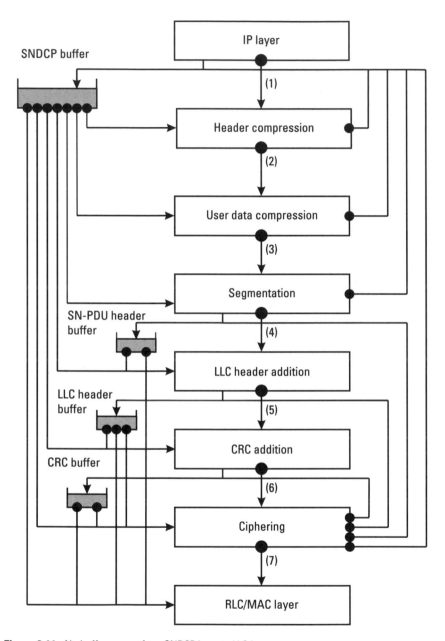

Figure 8.11 No buffer recopy from SNDCP layer to LLC layer.

Table 8.5
Advantages and Drawbacks of Solutions Implemented from SNDCP to LLC

	Buffer Recopy	**No Buffer Recopy**
Advantages	Easy implementation	Good performance in time
	Independence of layers between them	Low memory consumption
Drawbacks	Bad performance in time	Complex implementation
	High memory consumption	Dependence of layers between them

Example 1: Recopy from LLC to SNDCP

When the LLC layer receives a ciphered LLC PDU from the RLC/MAC layer, it performs the deciphering function and overrides the input buffer with the deciphered frame. If this LLC PDU corresponds to the first segment of N-PDU, then the SNDCP layer allocates a new buffer equal to the maximum length of an N-PDU. The N-PDU buffer is made up of several segments, each segment having a length equal to the data part of SN-PDU. Thus each received LLC PDU is copied in one segment of the N-PDU according to the SN-PDU received number without LLC header, CRC, and SN-PDU header. When all SN-PDUs have been received to be reassembled in one N-PDU, the SNDCP performs the user data decompression and copies the result of these operations into an uncompressed N-PDU buffer read by the IP layer. In this example, no IP header decompression is performed.

Figure 8.12 illustrates the first example with buffer recopy from the LLC layer to the SNDCP layer.

This solution is very simple to implement, being that the layers are independent of the others. But it implies a significant memory consumption, due to the fact that for each new N-PDU, the SNDCP allocates a buffer equal to the maximum length of an N-PDU.

Example 2: No Recopy from LLC to SNDCP

The beginning of the data processing in this example is identical to that of the previous example. When the LLC layer receives a frame carrying an SN-PDU, it provides the SNDCP layer with a pointer on the LLC data part carrying an SN-PDU. If it is the first segment of an N-PDU, the SNDCP layer allocates one buffer containing the list of received pointers on SN-PDUs from the LLC layer. Upon receipt of the last SN-PDU segment, the SNDCP

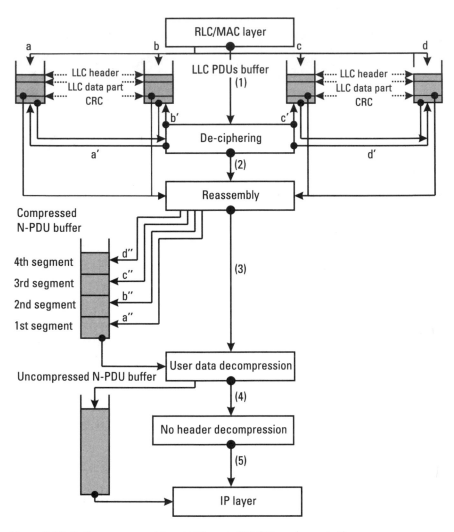

Figure 8.12 Buffer management from LLC layer to SNDCP layer (example 1).

layer reassembles the N-PDU by chaining the list of received pointers. Then the SNDCP layer sends the list of chained pointers on the SN-PDU header to the IP layer. In this example, no user data decompression and no header decompression are performed.

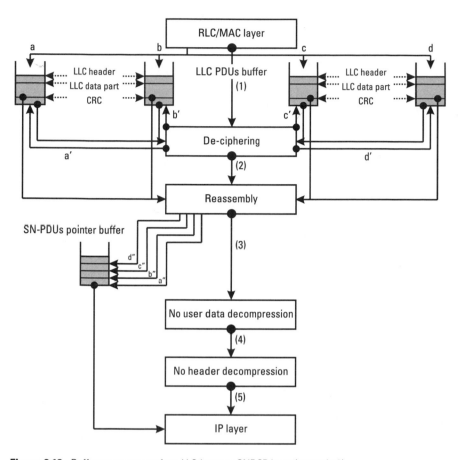

Figure 8.13 Buffer management from LLC layer to SNDCP layer (example 2).

Figure 8.13 illustrates the second example with no buffer recopy from the LLC layer to the SNDCP layer.

No memory recopy implies less memory consumption and a gain of time between each data processing event. But this solution is hard to implement because it is based on a list of chained pointers for each such event.

The comparison of these two solutions for buffer usage from the LLC layer to the SNDCP layer arrives at the same conclusion as that for buffer usage from the SNDCP layer to the LLC layer. It is provided in Table 8.6.

Table 8.6
Advantages and Drawbacks of Solutions Implemented from LLC to SNDCP

	Buffer Recopy	**No Buffer Recopy**
Advantages	Easy implementation	Good performance in time
	Independence of layers between them	Low memory consumption
Drawbacks	Bad performance in time	Complex implementation
	High memory consumption	Dependence of layers between them

8.2   GTP Layer for the User Plane

The GTP layer for the user plane (GTP-U) provides services for carrying user data packets between the GSNs in the GPRS backbone network. Packets from the MS or external data packet network are encapsulated in a packet GTP-U PDU (G-PDU) that consists of a GTP header plus a T-PDU within the GPRS backbone network. A T-PDU corresponds to an IP datagram and is the payload tunneled in the user plane GTP tunnel associated with the concerned PDP context. A GTP tunnel in the user plane is called a GTP-U tunnel and is created during PDP context activation or during PDP context modification procedures.

8.2.1   GTP-U Tunnel

A GTP-U tunnel is identified by a TEID, an IP address, and a UDP port number. The IP address and the UDP port number define a UDP/IP path, a connectionless path between two endpoints (i.e., SGSN or GGSN). The TEID identifies the tunnel endpoint in the receiving GTP-U protocol entity; it allows for the multiplexing and demultiplexing of GTP tunnels on a UDP/IP path between a given GSN-GSN pair.

8.2.2   Sequence Delivery

When the sequence delivery option is requested by the SGSN or the GGSN during the tunnel management procedures, the sending GTP-U protocol entity sends T-PDUs in sequence. A sequence number is associated with each in-sequence T-PDU and is present in the GTP header. The receiving GTP-U protocol entity checks the sequence delivery of T-PDUs and must reorder out-of-sequence T-PDUs if necessary.

8.3 IP Packet Sending Within GPRS PLMN

8.3.1 IP Packet Sending from MS to GGSN

Figure 8.14 illustrates a scenario for IP packet sending from MS to GGSN. The IP packet is received by the SNDCP layer on the MS side as an N-PDU. If the N-PDU size is longer than a given size N201, the SNDCP layer segments the N-PDU into several SN-PDUs. It then sends each SN-PDU to the LLC layer, which formats the PDU received into an LLC PDU frame either in UI frame for unacknowledged frame transfer or in I for acknowledged information transfer, depending on the reliability class for QoS. Next the LLC PDU is sent from the MS to the SGSN via the BSS by being encapsulated in RLC PDU on air interface and in BSSGP PDUs on Gb interface. The LLC layer within SGSN forwards to the SNDCP layer each LLC frame received into an SN-PDU. This layer reassembles SN-PDUs into an N-PDU if segmentation was performed on the MS side, and next forwards the N-

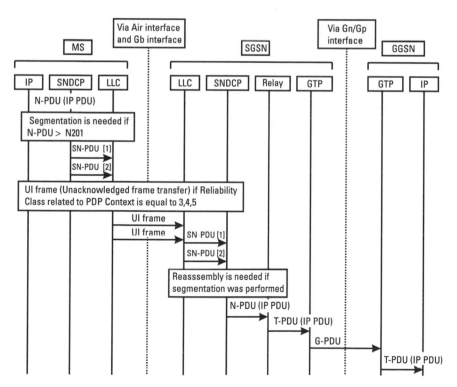

Figure 8.14 IP packet sending in uplink direction.

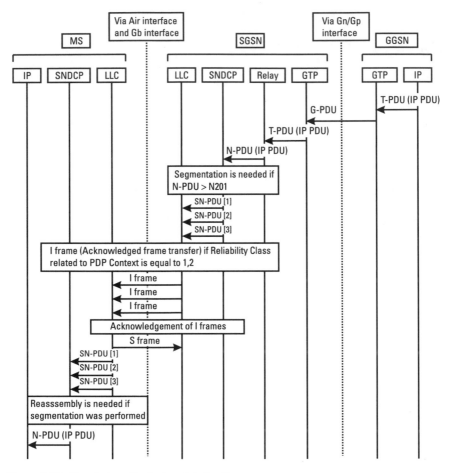

Figure 8.15 IP packet sending in downlink direction.

PDU to relay entity. The role of the latter is to adapt N-PDU format in T-PDU format and vice versa. G-PDUs are tunneled in GTP tunnels across the Gn or Gp interface between SGSN and GGSN. Next, the GTP layer in GGSN extracts the T-PDU from the G-PDU in order to send the T-PDU (IP PDU) to the IP layer.

Note that:

1. N201 is the maximum length of the information field in LLC frame.

2. IP fragmentation is processed on the Gn interface (or Gp interface) when the T-PDU size is larger than the *maximum transmission unit* (MTU) value. The latter represents the maximum datagram size that can be transmitted through the GPRS backbone network via Gn or Gp interface.

8.3.2 IP Packet Sending from GGSN to MS

Figure 8.15 illustrates a scenario for IP packet sending from a GGSN to an MS. This scenario follows the same scenario as IP sending from MS to GGSN but in the opposite direction.

8.4 Interworking with External Networks

A packet domain network that is a PLMN supporting GPRS is interconnected at the Gi interface to an external data packet network (IP networks). The Gi interface is located between the GGSN and the external data packet network. The GGSN is seen as an IP router by the external IP network. Figure 8.16 shows an IP network interworking.

The packet domain network supports interworking with networks based on the IP. It may also support interworking with networks based on the *point-to-point protocol* (PPP).

8.4.1 Interworking with PDNs Based on IP

The packet domain network provides either a direct transparent access to the Internet or a nontransparent access to the intranet/*Internet service provider* (ISP). In a direct transparent access, the MS IP address is allocated by the

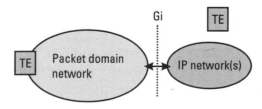

Figure 8.16 IP network interworking. (*From:* [1].)

GPRS PLMN operator within its addressing space. No user authentication or authorization process is performed for direct transparent access during the PDP context activation procedure. In a nontransparent access to an intranet/ISP, the MS IP address is allocated within the intranet/ISP addressing space. User authentication from a server (e.g. Radius, DHCP) belonging to the intranet/ISP is required during the PDP context activation procedure.

Figure 8.17 shows an interworking with PDN based on IP.

An L2 or PPP link exists between the *data terminal equipment* (DTE) and *mobile terminal* (MT) to encapsulate the IP packets only if the DTE and MT are two separate pieces of equipment making up the MS. The PPP layer implemented in the MS acts as a PPP server and the layer in the DTE acts as a PPP client.

Note that the PPP protocol consists of the following elements:

- A multiprotocol datagram encapsulation method;
- A LCP for establishing, configuring, and testing the data link;
- A family of *network control protocols* (NCPs) for configuring and establishing the different network-level protocols; the IPCP is implemented for IP configuration.

During a PDP context activation procedure, a PPP link is established between the DTE and MT in several phases:

- LCP configuration for the PPP connection (e.g., maximum size of the PDUs exchanged, activation or nonactivation of the authentication phase, activation or nonactivation of link-quality monitoring, compression of protocol, address and command fields);
- PAP or CHAP authentication phase (optional);

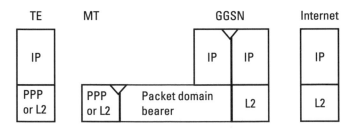

Figure 8.17 Interworking with PDN based on IP.

- IPCP negotiation (e.g., negotiation of IP parameters: Van Jacobson header compression, primary address of the DNS server, secondary address of the DNS server).

Figure 8.18 illustrates a PDP context activation for nontransparent access to ISP/intranet. The authentication request (PAP, CHAP) and IP configuration request are tunneled in *protocol configuration options* (PCOs) between the MT and the GGSN during the PDP context activation procedure. The GGSN will deduce from the APN the ISP address via a DNS server and the servers used for IP address allocation and for user authentication—as in the scenario below, where DHCP is used for host configuration and address allocation and RADIUS is used for authentication with the CHAP protocol.

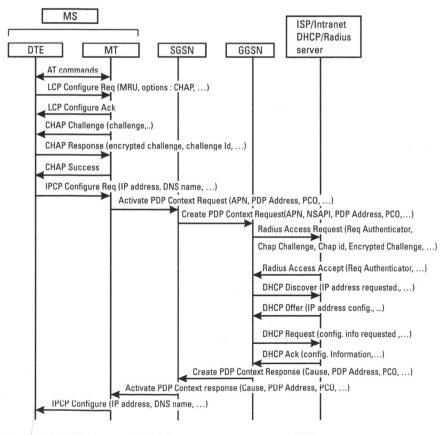

Figure 8.18 PDP context activation for non transparent access to ISP/intranet.

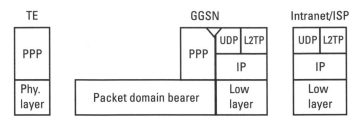

Figure 8.19 Interworking with PDNs based on PPP access.

8.4.2   Interworking with PDNs Based on PPP

A packet domain network may interwork with networks based on the PPP at Gi reference. Either PPP connections are terminated at GGSN or PPP frames are tunneled via the *layer-two tunneling protocol* (L2TP) at Gi reference.

Figure 8.19 shows an interworking with PDNs based on PPP access.

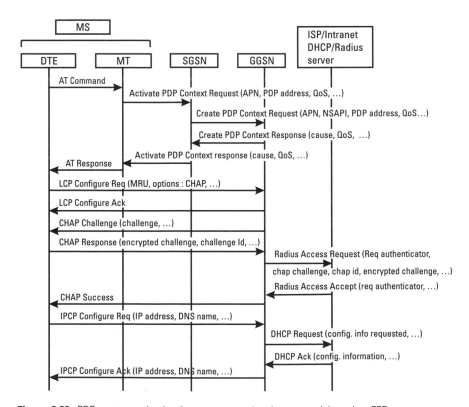

Figure 8.20 PDP context activation for access to packet data network based on PPP.

Figure 8.20 illustrates a PDP context activation for access to a PDN based on PPP. The MS IP address is allocated within the intranet/ISP addressing space. In the scenario below, we have assumed that the PDP address was given at the subscription. User authentication (PAP, CHAP) and retrieval of configuration parameters for connection to IP networks may be performed with Radius or DHCP server during PPP link establishment after the PDP context activation procedure.

Reference

[1]   3GPP TS 29.061 Interworking Between the Public Land Mobile Network (PLMN) Supporting Packet Based Services and Packet Data Networks (R99).

Selected Bibliography

3GPP TS 03.60 Service Description; Stage 2 (R97).

3GPP TS 4.64 Logical Link Control (LLC) Layer Specification (R99).

3GPP TS 04.65 Subnetwork Dependent Convergence Protocol (SNDCP) (R99).

3GPP TS 9.60 GPRS Tunneling Protocol (GTP) Across the Gn and Gp Interface (R97).

3GPP TS 23.060 Service Description; Stage 2 (R99).

3GPP TS 29.060 GPRS Tunneling Protocol (GTP) Across the Gn and Gp Interface (R99).

List of Acronyms

3GPP	Third-Generation Partnership Project
AB	access burst
ABM	asynchronous balanced mode
ADM	asynchronous disconnected mode
AC	address control
ACK	acknowledgment
ADC	analog-to-digital converter
AGC	automatic gain control
AGCH	access grant channel
AM	amplitude modulation
AoCC	advice of charge—charging
AoCI	advice of charge—Information
APN	Access Point Network
ARFC	absolute radio frequency channel
ARFCN	absolute radio frequency channel number
ARIB	Association of Radio Industries and Businesses
ARQ	automatic repeat request
AuC	authentication center

BC	bearer channel
BCCH	broadcast control channel
BCS	block check sequence
BDP	bandwidth delay product
BECN	backward explicit congestion notification
BER	bit error rate
BEP	bit error probability
BGIWP	barring of GPRS interworking profile
BH	block header
BLER	block error rate
BSC	base station controller
BSIC	base station identity code
BSN	block sequence number
BSS	base station system
BSSAP+	base station system application part+
BSSGP	base station system GPRS protocol
BTS	base transceiver station
BVC	BSSGP virtual connection
BVCI	BSSGP virtual connection identifier
CA	certification authority
CA	cell allocation
CBCH	cell broadcast channel
CCCH	common control channels
CDMA	code-division multiple access
CDPD	cellular digital packet data
CHAP	challenge handshake authentication protocol
CI	cell identifier
C/I	carrier-to-interference ratio
CIR	channel impulse response
CKSN	ciphering key sequence number

CLNP	connectionless network protocol
CLNS	connectionless network service
CN	core network
CONS	connection-oriented network service
CPM	continuous phase modulation
CQC	client query capability
CRC	cyclic redundancy check
CS	coding scheme
CU	cell update
CUG	closed user group
CV	countdown value
CWTS	China Wireless Telecommunication Standard Group
D	direction
DAC	digital-to-analog converter
dc	direct current
DCS1800	Digital Cellular System, GSM based on 1,800 MHz band
DISC	disconnect
DL	downlink
DLCI	data link connection identifier
DM	disconnected mode
DNS	domain name server
DRX	discontinuous reception
DSC	downlink signaling counter
DTE	data terminal equipment
DTM	dual transfer mode
E	extention
EDGE	Enhanced Data rates for Global Evolution
EGPRS	Enhanced General Packet Radio Service
EIR	equipment identity register
EMS	enhanced message service

ES/P	EGPRS supplementary/polling
ETSI	European Telecommunication Standards Institute
FACCH	fast associated control channel
FB	frequency correction burst
FBI	final block indicator
FCCH	frequency correction channel
FCS	frame check sequence
FDD	frequency-division duplex
FDMA	frequency-division multiple access
FECN	forward explicit congestion notification
FH	frame header
FH	frequency hopping
FN	frame number
FR	frame relay
FRMR	frame reject
FS	final segment
GERAN	GSM EDGE radio access network
GGSN	gateway GPRS support node
GMM	GPRS mobility management
GMSC	gateway mobile-service switching center
GMSK	Gaussian minimum shift keying
GPRS	General Packet Radio Service
GSM	Global System for Mobile Communications
GSMS	GPRS Short Message Service
GSN	GPRS support node
GT	global title
GTP	GPRS tunnelling protocol
HCS	header check sequence
HDLC	high-level data link control
HLR	home location register

HPLMN	home public land mobile network
HSCSD	high-speed circuit-switched data
HSN	hopping sequence number
HTML	HyperText Markup Language
HTx	hilly terrain propagation channel, with speed x km/hr
I	information
IF	intermediate frequency
IM	implementation margin
IMAP	Internet Message Access Protocol
IMC	Internet Mail Consortium
IMEI	international mobile equipment identity
IMGI	international mobile group identity
IMSI	international mobile subscriber identity
IN	intelligent network
IP	Internet Protocol
IP3	third-order intercept point
IPCP	Internet Protocol Control Protocol
IPLMN	interrogating PLMN
IP-M	Internet Protocol multicast
ISDN	Integrated Services Digital Network
ISI	intersymbol interference
ISL	input signal level
ISP	Internet service provider
LA	location area
LAC	location area code
LAI	location area identifier
LAPD	link access procedure on the D-channel
LCP	link control protocol
LI	length indicator
L2TP	layer-two tunneling protocol

LLC	logical link control
LLE	logical link entity
LMI	link management interface
LO	local oscillator
LSP	link selector parameter
M	more
MA	mobile allocation
MAC	medium access control
MAC	message authentication code
MAIO	mobile allocation index offset
MAP	mobile application part
ME	mobile equipment
MM	Mobility Management
MMS	multimedia messaging service
MS	mobile station
MSC	mobile-services switching center
MSK	minimum shift keying
MSISDN	mobile station ISDN number
MT	mobile terminal
MTP	message transfer part
MTU	maximum transmission unit
NAP	network access point
NB	normal burst
NCP	network control protocol
NER	nominal error rate
NF	noise factor
NM	network management
NMC	network management center
NNI	network-network interface
N-PDU	network protocol data unit

NS	network service
NSAPI	network service access point identifier
NSC	network service control
NSDU	network service data unit
NSE	network service entity
NSEI	network service entity identifier
NSS	network subsystem
NS-VC	network service virtual connection
NS-VCI	network service virtual connection identifier
NS-VL	network service virtual link
NS-VLI	network service virtual link identifier
NZIF	near-zero intermediate frequency
O&M	operation and maintenaince
OOS	origin offset suppression
OSR	oversampling ratio
OSS	operator-specific services
OTA	over the air
PA	power amplifier
PACCH	packet associate control channel
PAGCH	packet access grant channel
PAP	PPP authentication protocol
PAP	push access protocol
PBCCH	packet broadcast control channel
PC	power control
PCH	paging channel
PCCCH	packet common control channel
PCS1900	Personal Communication System, GSM based on 1,900-MHz band
PCU	packet control unit
PDA	personal digital assistant

PDC	Personal Digital Cellular
PDCH	packet data channel
PDTCH	packet data traffic channel
PDN	packet data network
PDP	packet data protocol
PDU	protocol data unit
PFI	packet flow identifier
PFM	packet flow management
PI	push initiator
PIM	personal information manager
PKI	public key infrastructure
PLL	phase locked loop
PLL	physical link layer
PLMN	public land mobile network
PNCH	packet notification channel
POP	Post Office Protocol
PO-TCP	PPG-originated TCP connection establishment method
PPCH	packet paging channel
PPG	push proxy gateway
ppm	parts per million
PPP	point-to-point protocol
PR	power reduction
PRACH	packet random access channel
PSI	packet system information
PSTN	Public Switched Telephone Network
PT	payload type
PTCCH	packet timing advance control channel
P-TMSI	packet-temporary mobile station identity
PTP	point-to-point
PVC	permanent virtual circuit

QoS	quality of service
R	retry
RA	routing area
RAC	Routing Area Code
RADIUS	Remote Authentication Dial-In User Service
RAx	rural area propagation channel, with speed x km/hr
RACH	random access channel
RAI	routing area identity
RAN	radio access network
RAS	remote access server
RBSN	reduced block sequence number
RDF	resource description framework
RF	radio frequency
RFL	reference frequency list
RLC	radio link control
RMS	root mean square
RNR	receiver not ready
RR	receiver ready
RR	radio resource
RRBP	relative reserved block period
RRM	radio resource management
RT	real time
RTI	radio transaction identifier
Rx	reception
RXLEV	received signal level measurement
RXQUAL	receive signal quality measurement
SA	service architecture
SABM	set asynchronous balanced mode
SACCH	slow associated control channel
SACK	selective acknowledgment

SAP	service access point
SAPI	service access point identifier
SATK	SIM application toolkit
SB	synchronization burst
SCCP	signaling connection control part
SCH	synchronization channel
SCP	service control point
SDCCH	stand-alone dedicated control channel
SDU	service data unit
SGSN	serving GPRS support node
SI	service indication
SI	stall indicator
SI	system information
SIM	subscriber identity module
SL	service loading
SM	session management
SMG	special mobile group
SMS	Short Message Service
SMS-GMSC	Short Message Service—gateway MSC
SMS-IWMSC	Short Message Service—Interworking MSC
SMTP	Simple Mail Transfer Protocol
SNDCP	subnetwork dependent convergence protocol
SN-PDU	SNDCP PDU
SNR	signal-to-noise ratio
SNS	subnetwork service
S/P	supplementary/polling
SRES	signed result
SS7	Signaling System No. 7
SVC	switched virtual circuit
T	terminal

TA	timing advance
TBF	temporary block flow
TCAP	transaction capabilities application part
TCH	traffic channel
TCP	Transmission Control Protocol
TDD	time-division duplex
TDMA	time-division multiple access
TE	terminal equipment
TEID	Tunnel Endpoint Identifier
TFI	temporary flow identity
TFT	traffic flow template
TI	TLLI indicator
TLLI	temporary link level identity
TLS	transport layer security
TMSI	temporary mobile subscriber identity
TOS	type of service
TPI	transport information items
TRAU	Transcoding and Adaptation Unit
TRX	tansceiver
TS	training sequence
TSC	training sequence code
TSG	technical specification group
TTA	Telecommunications Technology Association
TTC	Telecommunication Technology Committee
Tx	transmission
TUx	typical urban propagation channel, with speed x km/h
UDP	User Datagram Protocol
UI	unconfirmed information
UL	uplink
UNI	user-network interface

URL	Uniform Resource Locator
USF	uplink state flag
USIM	User Service Identity Module
USSD	unstructured supplementary service data
UWCC	Universal Wireless Communications Corporation
VC	virtual circuit
VLR	visitor location register
VPLMN	visited public land mobile network
WIM	wireless identity module
WWW	World Wide Web
XID	exchange identification
ZIF	zero intermediate frequency

About the Authors

Emmanuel Seurre is a system engineer in Alcatel's handset division. He has worked on all of the GSM circuit data transmission technologies and has experience with all of the mobile layers related to GPRS and EDGE at the system and standards levels.

Patrick Savelli has acquired baseband and RF system expertise at the mobile phone divisions of Alcatel and Mitsubishi Electric, especially on GPRS, EDGE, and UMTS, for which he has followed the evolutions in the standards groups.

Pierre-Jean Pietri has worked for Alcatel, on a system specification team, and followed the standardization process for EDGE and GERAN for the BSS side. He now works for STMicroelectronics on the development of technical solutions for GSM/EDGE handsets.

Index

Understanding GPS: Principles and Applications,
  Elliott D. Kaplan, editor

Understanding WAP: Wireless Applications, Devices, and Services,
  Marcel van der Heijden and Marcus Taylor, editors

Universal Wireless Personal Communications, Ramjee Prasad

WCDMA: Towards IP Mobility and Mobile Internet, Tero Ojanperä
  and Ramjee Prasad, editors

*Wireless Communications in Developing Countries: Cellular and
  Satellite Systems,* Rachael E. Schwartz

Wireless Intelligent Networking, Gerry Christensen,
  Paul G. Florack, and Robert Duncan

Wireless LAN Standards and Applications, Asunción Santamaría
  and Francisco J. López-Hernández, editors

Wireless Technician's Handbook, Andrew Miceli

For further information on these and other Artech House titles,
including previously considered out-of-print books now available
through our In-Print-Forever® (IPF®) program, contact:

Artech House   Artech House
685 Canton Street   46 Gillingham Street
Norwood, MA 02062   London SW1V 1AH UK
Phone: 781-769-9750   Phone: +44 (0)20 7596-8750
Fax: 781-769-6334   Fax: +44 (0)20 7630-0166
e-mail: artech@artechhouse.com   e-mail: artech-uk@artechhouse.com

Find us on the World Wide Web at:
www.artechhouse.com